Practical Field Guide to Grapegrowing and Vine Physiology

Practical Field Guide to Grapegrowing and Vine Physiology

Daniel Schuster, Andrea Paoletti
and Laura Bernini

Forewords by Michael Mondavi and Robert Mondavi, Jr.,
Mario Fregoni, and Richard Rowe

WINE APPRECIATION GUILD PRESS

San Francisco

Practical Field Guide to Grapegrowing and Vine Physiology

Text copyright © 2018 Daniel Schuster, Andrea Paoletti, Laura Bernini

No part of this book may be reproduced or transmitted in any form or by any means, electronic or mechanical, including photocopying, recording, or by any information storage and retrieval system, without permission in writing from the copyright holder.

Wine Appreciation Guild an imprint of
Board and Bench Publishing
www.boardandbench.com

Copy editing by Judith Chien

Design and composition by Publishers' Design and Production Services, Inc.

ISBNs

978-1-935879-31-2 (print)
978-1-935879-06-0 (ePub)

Library of Congress Cataloging-in-publication is on file with the Library of Congress

Although all reasonable care has been taken in the preparation of this book, neither the author nor the publisher can accept liability for any consequences arising from the information contained herein, or from use thereof.

Contents

Foreword by Michael Mondavi and Robert Mondavi, Jr.   vii

Foreword by Richard Rowe   ix

Foreword by Mario Fregoni   xi

Introduction and Acknowledgements   xiii

About the Authors   xvi

Chapter 1	Viticulture	1
	Historical perspective	1
	Grape vine adaptations to a site, climate and soils.	7
	Systems of viticulture—industrial, sustainable, organic, bio-grow and bio-dynamic.	10
Chapter 2	Winter	15
	Vine dormancy, morphology and physiology of the roots.	15
	Root morphology and water/nutrient uptake.	16
	Principal pruning and canopy training systems.	20
	Quadrilateral cane or cordon-spur pruning system with VSP or V-shape canopy in winter.	32
	Rootstocks for different soils and climates.	35
	Pathogenic fungi and bacteria affecting vines, pruning sanitation.	40
Chapter 3	Spring	45
	Budburst and early shoot growth.	45
	Bud fruitfulness, flowering and fruit set.	53
	Canopy construction, canopy quality and optimal size.	58

Chapter 4	Summer		65
	Berry and seed growth from fruit set to veraison.		65
	Respiration rates, water and nitrogen usage by vines.		66
	Photosynthesis, carbohydrates, crop levels and phenols.		70
	Berry weight, volume and composition changes from veraison to harvest.		73
	Vigour control, irrigation and dry farming.		74
Chapter 5	Autumn		79
	Yield estimates and grape harvest maturity indexes.		79
	Mechanical or hand harvest.		87
	Post-harvest vine management.		88
Appendix 1	Bibliography, Literature, References and Further Reading		97
Appendix 2	List of Tables, Figures and Illustrations		101
	Tables		102
	Figures		102
	Photo illustrations		102
Index			105

Foreword

by Michael Mondavi and Robert Mondavi, Jr.

I have known Danny Schuster for more than two decades and have had the pleasure of walking through our family vineyards with him and my son many times—each visit is a new learning experience. When I am later recalling the information Danny shared with us, this book provides an invaluable reference to better understand the balance of the soil, the climate and the vine.

The more I continue to learn about winemaking and viticulture, the more I realize there is still so much more to know. Danny Schuster has taken me a huge step forward in my quest for knowledge about the mystery of the grape vine, the soil and climates that impact them and Danny's commonsense approach is understandable and, to me, one of the most helpful books written on viticulture.

—Michael Mondavi

When reading Danny Schuster's book I was captivated by the beauty of the vine, and how we as winemakers harness its bounty, shaping liquid into a form representing the earth, sun and sky. The elegance of Danny's work is how real data and research are melded together, extracting the vital essence of how and why the vines will yield us the highest gifts when understood and harnessed.

Traveling through the Napa Valley, Tuscany or Bordeaux, we have seen perfectly manicured vines and pristine farms and know that we are creating the best wines of our time. Yet, I believe that every vintner can become a student of the vine, find a new passion and enhance their wines through better vineyard practices. Through Danny's eyes, I see that there is more work to be done. What Danny outlines in his book are readily attainable and quantifiable methods that can unlock the next evolution in vineyards and winemaking that will bring a new era of wine to life.

This book has reaffirmed my belief in the importance of keeping the magic of wine alive, while unlocking the challenging mystery behind it. I hope you will enjoy this book and keep it nearby for reference.

—Rob Mondavi, Jr.

Foreword

by Richard Rowe

Professor Emeritus, Horticultural Science Lincoln University.
B. Ag. Sci. (Melbourne Univ.), M. Hort.(Univ. Calif. Davis),
PhD (Univ. Calif. Davis).

There have been many thousands of words written for the benefit of the consumers of wine, describing the myriad of wine labels and wine styles available to the wine drinkers throughout the world. Many words have been dedicated to ranking wines according to their perceived quality and flavour characteristics, to aid the consumer in selecting what to drink. Far less has been written about the detailed processes carried out in the vineyards to produce the grapes from which the wines are made.

This book brings together the extensive knowledge over the distinguished careers of the author(s) in teaching, practising viticulture and research in most areas of the world in which wine is made. It logically sets out in chronological order the many decisions and steps necessary from establishing of the vineyards to annual maintenance of the vines and overcoming the many natural hazards and local conditions that must be considered to bring the annual vintage to maturity and sustain yields over the lifetime of the vineyard, to achieve the best outcome of which any particular vineyard is capable. Each step is prefixed by the aim of the procedure, which is being attempted in an effort to overcome or achieve specific goals.

This book should be the foundation for every trainee's and dare I say old viticulturist's work plan, a check list against which the routine vineyard operations are planned throughout the year. It includes not only details of steps to be taken, but also provides methods derived from research and experience to accurately measure how well the desired objective has been achieved, adding precision to the operation of a vineyard and a predictor of outcomes, whether it be sustained yield, grape quality or a combination of both. In particular it emphasizes achieving a balance between vegetative growth (leaf and stem) and reproduction (seed and fruit growth), appropriate to the specific terroir of the vineyard. This balance is critical in achieving long-term yield and quality in wine grape production.

What the author(s) bring to the fore is that no one universal method fits all vineyard situations; however, there are biological principles which are peculiar to the physiology of the grape vine and against which the methods adapted in the vineyard can be judged.

I strongly recommend this book for any student of wine grape growing, whether training for a professional career in viticulture or an amateur who wishes to understand more about the processes followed in the production of grapes that go into the wines they drink and which bring them pleasure. It is a truism that the quality of fine wine starts in the vineyard. It is the role of the winemaker to ensure that the processes of wine making do not destroy the unique characteristics and quality which each specific terroir is capable of producing.

Richard N. Rowe,
Christchurch, New Zealand,
November 2013.

Foreword

by Mario Fregoni

Professor Emeritus Faculty of Viticulture, University of Piacenza
Honorary President of I.O.V. (International Organization du Vin), Paris

"Canopy management is vine physiology"
(Prof. M. Fregoni).

In my professional life as a professor of viticulture, I always favoured vine physiology. I remember that during a course of study of hail damage, I decided to study vine physiology instead, because hail damage is after all reflected in consequences to vine physiology; thermal stress, physical damage to shoots, leaves or grapes and nutritional damage to vines. Mechanical harvesting has physiological consequences also; in fact, every choice made by the grower be it rootstock, variety, pruning or canopy management, brings out physiological responses in vines.

This book is a scientific contribution to vine physiology and at the same time a practical guide to grape growing, because it contains professional experience of 3 international experts, working in a number of diverse countries such as Italy, New Zealand, California and Turkey. The text is a global overview of vine physiology, presented in 4 seasons of the year. The vines are studied not only in the main and well known aspects of the annual canopy, but also in their perennial parts and roots, as they both have physiological interaction that is important, even if it is very complicated. As I wrote before, a vine has two brains: apex and roots, both set up by meristematic cells very similar to stem cells in humans. Canopy growth and root balance are thus guided by hormonal action in young cells of the shoot apex and those in the roots.

In respect of roots, the authors deal with rootstock choices, soil environment, deep and radial root growth, root metabolism or water and mineral uptake, all of which affect the most important aspects of vine's life. Likewise, the canopy photosynthesis, transpiration and respiration are covered as a physiological foundation for phenologic phases of shoot growth, flowering, fruit set and berry maturation. Of special interest is the part dealing with the effects of the ratio of young to old leaves, the authors finding that the ratio of 75:25% up to flowering/fruit set, 50:50% during green berry growth, 25: 75% at veraison and finally 10:90% ratio during final maturation to harvest the best balance. Clearly, the very young, mature and the very old leaves have different but complimentary roles, such as the nitrate driven growth and accumulation of growth hormones in young, high rates of photosynthesis of carbohydrates in mature and synthesis of phenol compounds and accumulation of dormancy hormones in the old leaves.

The authors also show useful data on total leaf exposure among the principal pruning, vine spacing and canopy training systems. Also useful are the conversion rates of the weight increase coefficient of clusters from veraison to harvest weight of the same grapes at ripeness. Depending on the variety, this parameter fluctuates between 1.4 and 2.6; it is thus possible to accurately predict at hard seed stage of veraison the final weight of grapes at harvest per vine and organise the green harvest accordingly.

The book's contribution is much greater than this foreword can describe, its greatest value originating from the background of the authors, each of whom deals with vineyard management in all seasons and a great range of terroirs and the physiological interpretation for each. This well illustrated text should be important for all viticulturists, researchers, students at universities and wine schools around the world. It will be important to have it translated into Italian also. The book, like the authors, is international, and covers experiences of many countries of advanced viticulture around the world. An international acknowledgement (International Organization du Vin) is the recognition that this writer and a friend of the authors expresses in his foreword.

Mario Fregoni,
Piacenza, Italy,
February 2014.

Introduction and Acknowledgements

The aim of this book is to provide a guide to vine physiology and its practical application in the production of grapes for fine wines. The material presented here should provide growers with a working understanding of all vineyard processes described and where considered useful, the text is illustrated with diagrams, tables, drawings and photographs.

In an effort to retain focus on the practical, field application of grape growing advice, the authors limited the coverage of a number of related topics, included in greater detail and region-specific manner in other viticulture textbooks and technical manuals. For example, the impact of irrigation, fertilization, disease control is included, whilst the often region- /machinery-specific mode of operations is not. Likewise, the basic horticultural, agronomic or mechanization information has been greatly reduced also. Important as these topics are, they are usually well covered in viticulture literature, which can be found amongst the references, literature and further reading (Appendix 1).

The progression of topics in the chapters that follow is arranged by chronology of the seasons, as experienced by vine grower and the vines, irrespective of climatic zone of production or any region specific pruning/training system used or the grape variety grown.

One of the authors' key aims was to remain inclusive of all the varied viticultural systems in use at the present time, respecting both the traditional and modern approaches to growing vines. We expect this book will have special value in many areas of the 'New World' where grape growing tradition is more recent and experimentation or innovation is all important. It should also have value for growers in the well established, classical regions of the 'Old World' for all those that wish to improve their understanding of adaptations and natural tendencies of vines, which are mirrored in vine physiology and expressed in the local grape growing traditions for many centuries past.

It has been a fortunate coincidence that the authors of this book accumulated an extensive viticultural and wine making experience in many parts of the wine world from Turkey, Georgia to Eastern Europe, Germany, France, Italy, California, Chile, South Africa and Australasia.

Their long-term shared passion for both, vines and wine, resulted in this book.

"There is no real difference between tradition and innovation—only time" (Warren Winiarski, pioneer vigneron of Napa Valley, California).

The collective knowledge of mankind on any topic of human endeavour, such as grape growing, is the result of contributions by many countless individuals. In the case of cultivating grape vines, a millennia-old effort, the accumulated experience and wisdom comes to us via many generations of growers and

ILLUSTRATION 1.1 Wild vine growing in a tree.

winemakers who came before us. The authors were fortunate to benefit from advice and draw on a far more extensive pool of knowledge from others, before and during the preparation of this book. Amongst them, a special thanks and gratitude goes to

Profs. A. Carbonneau, E. Peynaud and Glories of Bordeaux Univ.

Prof. M. Fregoni of Piacenza Univ.

Profs. Boselli, L. Mugnai and Mattii of Florence Univ.,

Prof R. Rowe and Dr. D. Jackson of Lincoln Univ.,

Profs. H. Becker and Kiefer of Geisenheim Univ.,

Dr. L. Raspini and Axel Heinz of Tenuta del Ornellaia in Bolgheri,

W. Winiarski of Stag's Leap Wine Cellars in Napa Valley,

Dr. P. Dry of Adelaide Univ.,

Dr. R. Smart of Tasmania and

Profs. J. Morrison, Meredith and R. Bolton of Univ. California Davis,

Profs. P. Galet of Montpellier Institute and E. Peynaud of Bordeaux University,

Dr. L. Morton, Vit. Consultant Napa Valley,

Dr. John Hunt of Agrimm Technologies in New Zealand,

Michael Silacci of Opus 1 Winery in Napa Valley,

Thomas Duroux of Château Palmer in Medoc,

Bernard Moueix of Château Taillefer in Pomerol,

Olivier Leflaive from Côte d'Or, France and

Robert Mondavi of R. Mondavi Winery in Napa Valley.

For their professional skills our thanks go to our publishers, editors and translators, and to all the photographers that provided illustrations used. Not least of all, our gratitude go to our families for their patience and support over many years and vintages.

> Danny Schuster, Andrea Paoletti and Laura Bernini.

"If growing grapes and making wine was left to God, all we would end up with is vinegar"

> (Prof. Emile Peynaud, Bordeaux University, France).

About the Authors

Daniel Schuster's nearly 50 years in the cellars and vineyards of some of world's most renowned wineries, including his own, Daniel Schuster Wines of New Zealand, have made him one of the world most respected living viticulturalists. He is featured in a permanent exhibition at the National Museum of American History in the Smithsonian Institution of Washington D.C. for his contribution to the evolution of the modern Californian wine industry. Schuster is the co-author of Grape Growing and Wine Making for Cool Climates.

After a decade of distinguished vineyard management work in the Italian corporate wine world, in 1999 **Laura Bernini** became a freelance viticultural agronomist working in Tuscany and Umbria.

Andrea Paoletti was the longtime manager of Antinori Chianti Classico com-panies, before becoming a freelance consultant, working in vineyard locations from US, Hungary and Italy to Turkey, Montenegro and Georgia.

CHAPTER 1

Viticulture

Historical perspective

The species of grape vines (Vitis vinifera L.) belongs to a large family of climbing, forest plants (Ampelidaceae) that evolved some 70 million years ago in the moderate climate regions of Northern China and Trans-Caucasus Valleys in Asia. The natural habitats of wild grape vine species (V. silvestris, V. alba etc.) forced the development of genetically stable natural vine tendencies, such as adaptations to periodic drought conditions, sensitivity to light length and quality (phototrophy) and apically dominant habit of growth. Wild vines, like all other perennial, woody lianas, had to develop the capacity to produce carbohydrate surpluses, store them and release or allocate these in form of growth energy in a preferential way to favour either rapid elongation of dominant shoots, expansion of leaf area or to produce more fruit, thus seeds.

The natural (endemic) distribution and spread of species of wild vines (Vitis), caused by periodic changes of climates and habitats in regions of their origin over the past millennia, extended their presence into many diverse regions of the world in Asia, North America, the Caribbean and Europe. The present complexity of genetic relationship between the various species of grape vines is due not only to the introduction of cultivation and selections of varieties by man, but also due to a long-term interaction of wild species with each other (inter-specific hybridization) and from adaptations of the grape vines to new habitats of their changing distribution.

Naturally occurring mutations in specific locations of either wild or cultivated vines have resulted in further, more or less stable changes to vines morphology, attesting to ongoing adaptations to site-specific conditions. The task facing an ampelographer attempting to identify or classify vines by their morphology alone is, to separate characteristics that are due to cultivation, site adaptations from natural variability within the specie and all of its varietal forms. In botany, several genetic marker characteristics are required for separation at the sub-specific, varietal level of classification. In reality, leaf shape or berry color alone are also insufficient at varietal level for grape vines; thus terms like sport, form or cultivar would appear more correct than variety, where Cabernet or Riesling are concerned. As grape vines are propagated by vegetative means (cuttings, grafts, buds, green stems/tips etc.) the genetic instability of grape varieties, if grown from seed, is of no practical interest to the grape grower.

The selection of seedlings, in an effort to hybridise rootstocks or new inter-specific varieties, is another matter altogether and does not fall within the scope of this book. The more recent studies of the DNA of the various groups of grape vine varieties are of more interest, as they offer the best opportunity to map out the relationships and genetic origins of modern varieties, often grown under different local names or synonyms.

The origins of viticulture are attributed to Asia, the most likely region being the temperate climate zone of Trans-Caucasus, where the semi-nomadic methods of mixed farming are said to originate. Rudimentary cultivation of vines in semi-nomadic fashion may be as old as 15,000 BC, when plots of local fruit and food producing plants were established in close proximity to water sources and within the perimeter of the family or tribal controlled territory.

TABLE 1.1 Evolution of genus Vitis. (L.), after P. Galet.

The regions between the Black and Caspian seas proved especially favorable for early domestication of wild grape vines. Once under cultivation and regular winter pruning regime, the behavior of wild vines changed to increase fruitfulness at the expense of growth vigor, cultivated vines becoming self pollinating and producing larger clusters of more numerous berries. The original selections of cuttings of wild vines from their habitat was perhaps based on choosing the most productive plants at first. Later on, after being able to turn stored grapes into wine, the early grape growers may well have added 'wine quality' parameters (color/taste etc.) among their selection criteria for finding the most desirable mother vines.

From 5,000 BC onward, the cultivation of vines and the art of making wine progressed from Asia to Middle-Eastern regions, changing the semi-nomadic farming format into more settled farming practices on a larger scale. Assyrians, Persians and Egyptians, as well as the growing city states of Asia Minor, enjoyed grapes for eating and wines to drink or for medicinal use for millennia before Europe. Phoenicians, the early navigators and merchants of the Mediterranean, were known to trade in wine from 1,000 BC onward. At first, their wines would have originated from Egypt and central regions of Asia Minor; with time, vineyards were also planted on Crete, Cyprus and other islands, finally reaching as far as Carthage in North Africa.

It is reasonable to assume that the tribes of early Greece brought with them vine cuttings and the knowledge of making wine during their migrations from Asia Minor, and also that the Greek settlers brought viticulture to Sicily, Puglia and other parts of southern Italy, thus becoming instructors of viticulture to Etruscans and later on to Romans as well. The rapid expansion of vineyards

ILLUSTRATION 1.2 Coast of ancient Thracia with Gobelet vines, Turkey.

throughout Italy occurred during the early centuries of the Roman Empire, reaching the northern region of the Alps by the first century BC.

The oldest vineyards of France are those of Massilia (Marseilles), planted by the Greeks as early as the 5th century BC. Under the Romans, during the 1st century AD, viticulture reached all of the occupied provinces in central, eastern and northern Europe, as well as Spain. Grape growers in the Rhône, Mosel and the Rhine valleys, as well as Bordeaux, Burgundy and Champagne regions, recorded their first vintages at that time. By this stage, viticulture was well established in most climatically suitable areas of the Roman Empire, reaching as far as the southern and southwest provinces of Britannia (England).

The fifth century AD witnessed the collapse of the western parts of the Roman Empire, by which time wine production was well mastered by native populations in all modern viticultural areas of Europe. From there on, the region-specific development, including the grape varieties grown and type of wines made, occurred in ways that reflected the local climate, soils, socioeconomic and religious factors that continue to play their role to this day. More recently, the colonization of many parts of the 'New World' caused the spread of viticulture into all of the present-day wine producing countries. In America, vines from Europe were first introduced by Cortez in 1524, though native vines (V. labrusca, V. riparia, V. rupestris etc.) were already growing wild. In 1652, Jan van Riebeeck brought grape vine cuttings into the newly established province of Cape of Good Hope in South Africa. The Australian and New Zealand wine industry owes much to early efforts by James Busby, a British botanist and wine enthusiast, who introduced European vines to both colonies in the early decades of the 19th century.

Selection of grape varieties for different climatic zones.

Climate is one of the major factors determining where grapes can be grown and the potential wine quality produced from those grapes. Seasonal variations apart, the long-term climatic conditions limit the world's wine production into so-called 'temperate zone' within the 30–50 degrees latitude in the Northern Hemisphere and 30–40 in the Southern Hemisphere. It should be noted that not only the total amount of light, heat or rainfall are of importance, but also their distribution during the annual cycle, especially during the vine's growing season.

Close to the equator, where the mean temperature exceeds 20 degrees C, the winters are mild and leaf fall/dormancy do not occur or only partially so, the vines become non-productive. On the other hand, if the mean temperature drops below 10 degrees C, the summers will be too short and winters too severe. During short summers vines will have insufficient time to develop full canopies to ripe n their fruit, whilst extremes of cold during winter will cause permanent damage to roots and other perennial parts of the vine. The temperature range seen in the temperate climate zones, a mean temperature in the 10–20 degrees C range, are the two extremes within which grape vines (V. vinifera) will grow; however, not all of the 10,000 or so grape varieties known will produce quality grapes required for fine wine.

The initial, broad selection of varieties and resulting wine types can simply be based on calculations of total heat accumulation during the growing season, from bud burst to harvest. Using the 10 degree C (or 50 degrees F) base required for active vine growth, the UC Davis established a group of five climatic categories, from cold to hot climates. Each of these has a group of recommended varieties. This broad classification is sufficient for separating major wine categories, such as the light bodied, aromatic whites (Cat. I), from full-bodied red wines (Cat. III) to the industrial, bulk quality wines (Cat. IV) or fortified wines (Cat. V).

The requirement for more precise classification amongst the grape varieties for production of fine wines at the lower end of the Davis scale in categories I and II appears obvious and is well supported by the fact that most of the world's best table wines are produced at the lower temperature range, thus in regions accumulating less than 1,500 degree days of useful heat total. Here the vines grow without excessive heat stress, under greater diurnal temperature variations and slower maturation rates in the autumn—all of which results in improved balance of sugars/acids/pH, as well as enhanced flavour/aroma concentration and complexity of character of the resulting wine.

It can be argued that the warmer climates produce greater yields and experience reduced risk of unfavorable seasons, faced by the grape growers in cool to moderate climate regions. Spring frosts, increased incidence of fungal diseases and imperfect flowering in summer or delayed fruit ripening in the autumn are all hazards associated largely with cool climate areas. Nevertheless, wine production remains economically viable here because in favorable vintages the wine quality justifies premium prices that more than make up for the lower yields obtained and potential hazards of the climate.

Heat units calculations have shown to be a simple and useful guide for determining the general suitability of sites and varieties within a limited geographical

TABLE 1.2 World distribution of viticulture (after Jackson & Schuster 2001).

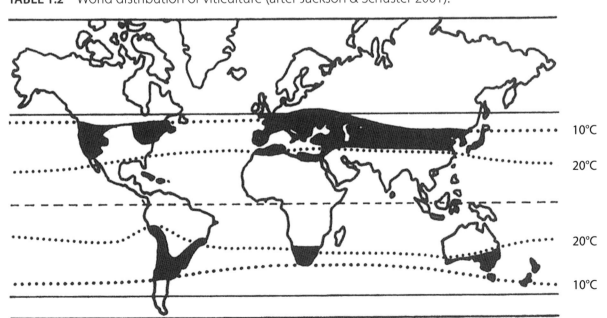

area of more or less uniform climatic patterns. On the other hand, heat units alone are of considerably less value when dealing with regions that are far apart, those of different latitude, altitude or climatic patterns, for example maritime versus continental climates. It will be observed that increased length of growing season or greater light intensity compensates for lower heat summation totals.

Increased altitude of a site or less favorable exposure to sun or wind also reduce the light or heat received by vines. It has been proposed that the mean temperature of the warmest month is comparable to the calculations based on total heat summation, as long as both values being compared are measured within the same climatic zone.

The length of the growing season is largely dependent on latitude, and studies carried out would suggest that latitude may well prove to be a more accurate indicator of a variety's suitability to given site than the degree days heat summation alone. The latest guide is the latitude and temperature adjusted index, or (LTI) calculated as; LTI = mean temperature of the warmest month × 60 (minus latitude).

Table 1.3 below provides a selection of key varieties showing both groupings, those based on heat summation calculations (degree days) and the LTI system.

Another potentially limiting factor of climate affecting grape production is the severity of winter chilling. In the extreme, as experienced in parts of continental Europe, Asia or North America, winter temperatures of minus 15 degrees C or lower tend to cause severe vine damage and if frequent, vine death. It has

TABLE 1.3 Selection of recommended varieties for different climatic zones (after Jackson & Shuster)

Scale of UC Davis degree days.	LTI scale after Lincoln Univ.	Selection of recommended varieties
Category I—cold less than 1390 C.	Group 1-A LTI less than 190	Siegerrebe, Ortega, Madelaine × Angevine 7672, Seyval blanc, Reichensteiner, Müller-Thurgau, Huxelrebe Bacchus and others.
Category II—cool less than 1590 C.		Pinot gris, Pinot blanc, Chenin blanc, Chasselas, Kerner, Traminer, Sylvaner, Faberrebe, Sauvignon blanc, Aligoté, Scheurebe, Auxerrois, Aligote and group of varieties for 'traditional method' sparkling wine, Chardonnay, Pinot noir and Pinot meunier.
Category III—moderate less than 1790 C.	Group I-B LTI 190–270	Riesling, Pinot noir, Semillon, Chardonnay, Vermentino, Sauvignon blanc, Veltliner, Muscat blanc and others.
Category IV—warm less than 1990 C.	Group I-C LTI 270–380	Cabernet sauvignon, Cabernet franc, Merlot, Syrah, Petit Verdot, Malbec, Marsanne, Rousanne, Viognier, Sangiovese, Nebbiolo, Teroldego, Mourvedre and others.
Category V—hot less than 2,200 C.	Group II LTI 380 plus	Grenache, Syrah, Tannat, Carignan, Tempranillo, Cinsault, Zinfandel (Primitivo), Mourvedre, Verdicchio, Nero D'Avola and others.

(After Jackson D. & Schuster D. 2001)

ILLUSTRATION 1.3 Winter landscape in cool climate region.

to be noted that grape varieties of northern Europe are more resistant to winter chill than those of the southern regions; however, none can withstand the cold temperatures indicated above. Some of the native American species of vines (V. berlandieri) or those of Asia (V. siberica) show greater resistance to winter cold, but so far, all attempts to introduce such resistance into V. vinifera via interspecific hybridization, as in Franco-American or V. vinifera × V. siberica hybrids, failed in the terms of wine quality.

At the other end of the temperature scale, experienced in hot regions at lower latitudes, the absence of full dormancy caused by high soil and air temperatures of ambient, above 10 degrees C and/or an insufficient chilling period in the root zone (60–90 days at less than 6–8 degrees C required), results in the lack of fruitfulness, which makes viticulture in sub-tropical or tropical regions a difficult and uneconomic proposition.

Grape vine adaptations to a site, climate and soils.

The genetically carried, long-term tendencies of grape vines caused by adaptations to their original habitats in Asia have, after vines were transported to their new environments by civilizations of man, been supplemented by further adaptations and/or mutations of the original wild vines into modern varieties known today.

Apart from the strongest tendencies that are shared by all grape varieties (drought responses, light sensitivity and apical dominance), other, more subtle and less stable changes to vine morphology can occur if the variety is planted in regions of significantly changed terroir. Vine growth vigor, size and the shape of the leaf, type of aroma/flavour profile and their concentrations all attest to the ability of vines to adapt to different climates and sites.

ILLUSTRATION 1.4 Summer landscape in hot climate region.

The soil's density and structural components of the sub-soil profile are of special interest here, as they shape the root morphology, which in turn governs the water/nutrient uptake and vine vigor. Since the advent of grafting V. vinifera onto American rootstock species, adaptations of the European vines to soil conditions alone have become more difficult to study outside of research institutions or vineyards still planted on their own roots, such as those of Chile, South Australia, southern parts of New Zealand and Washington State.

Adaptations to pruning systems, canopy training and those imposed by the seasonal and long-term changes to climatic conditions can be more obvious. The growers observe the impact of cold, less favorable seasons from the effects of hot/dry vintages. Modifications of the various pruning systems and of canopy/crop management will also be reflected in per bud fruitfulness and fruit composition. Certain varieties, especially those of spontaneous, inter-specific hybridization origin, will demonstrate a greater degree of variability, especially those that are prone to mutations or grown in regions of drastically different conditions from those of their origin, over a long period of time.

These region-adapted forms, bio-types or clones of vine varieties will often be found under various, region-specific names (synonyms), despite their close relationship and their identical genetic footprint. Sangiovese in Italy, Pinot and Syrah in France or Riesling in Germany are good examples of this trend. The fact that by DNA alone, the Sangiovese in Chianti, Brunello in Montalcino, Prugnolo in Montelpuciano or Nielluccio in Corsica belong to the same genetic group may be obvious to a trained ampelographer but, perhaps less so to the individual grape grower and the consumer of the resulting wines.

It has been estimated that close to 3,000 grape varieties are under commercial cultivation at present, yet, as the work of Prof. Bandinelli in Florence University or that of Prof. Galet at the Montpellier Research Institute in France shows, there could be as many as another 10,000 as yet unnamed, potential varieties in Italy alone. A comparable or even greater number is likely to exist in the less well explored regions of Balkans, Asia Minor and central parts of southern Asia, including ancient regions like Georgia or Trans-Caucasus. Many of these anonymous and long abandoned varieties may not be of commercial value however; some may have the potential for fine wines, whilst others offer a great pool of genetic material, no longer expressed in the modern varieties.

The last, but not least of adaptations forced on vines during the past century worldwide, are the rapidly growing industrial systems of large scale grape production, accompanied by greatly increased introduction of agrochemicals, pesticides, herbicides and industrial fertilizers into the vineyard environment. The responses of vines to general pollution and presence of toxins in the soil have proved to have negative impact on vine's own immunology, soil fertility and biology or bio-diversity in the vineyard environment, all of which is reflected in the composition of the wines made. The growing interest in and the growth of practical applications of the sustainable, organic and bio-dynamic forms of viticulture offer a long-term alternative to the industrial approach. Considering the ongoing adaptations of grape vines to new regions and to rapidly changing climatic conditions in many parts of the viticultural world, there can be no doubt that ongoing field selections of grape varieties, as well as re-introductions of the varieties from the past, may bring significant benefits to vintages yet to come.

ILLUSTRATION 1.5 Vine response to high soil salinity, Calabria.

Systems of viticulture—industrial, sustainable, organic, bio-grow and bio-dynamic.

The spread of viticulture into the New World and the rapid rise of demand for table wine in many new markets outside of Europe since the 1970s, was followed by major expansion of the vineyard area planted. The mechanization of vineyard operation and the advent of the widespread use of irrigation, fertilization, agrochemicals including herbicides in most 'New World' regions, became the key tools in industrialization of growing grapes.

Whilst the artisanal, multi-crop way of growing grapes in the traditional regions of Europe declined on account of reduced consumption of wine, production costs and labour shortages, the industrial production in eastern Europe, Russia, Australia, South Africa, Chile, Argentina and California increased at a great pace. Large, mechanized and automated wine cellars were built and machine harvesting was introduced, reducing significantly the production costs, favouring large scale vineyard operations and production of sound, if basic table wines.

There is no argument that the availability and overall quality of these industrial wines was improved by the introduction of modern technology, albeit at the price of individuality and character. The key inputs have changed from hand labour to investment in a large capital, machinery and energy. Whilst the uniformity and drinkability of these varietal, mass blended wines increased, it was achieved at the expense of environmental damage and loss of typicity of the region-specific character of the resulting wines. The growing acceptance of mass-produced wines by the new markets provided an outlet for the growing volume

ILLUSTRATION 1.6 Terraced old Nerello vines on Etna, Sicily.

of grapes and wine made during the prosperous two decades since the 1980s. The globalization of trade and ease of transport also made it possible for the traditional European producers of wine, like Italy and Spain, to catch up and even surpass a once major exporter like France, thus limiting the economic effects of falling domestic wine consumption.

From a viticultural point of view, the rise of the industrial approach to grape growing made it possible to plant vineyards in increasingly hot, arid climates (Australia, South Africa and California), as well as to establish large scale wine production in countries with small populations, like New Zealand. The increased capital cost required by the industrial system of farming forced the replacement of vine grower 'vigneron' with the corporate investor, whose short-term profit and market strategies replaced the need for longer-term sustainability and production of region- and site-specific wines, that express best the unique terroir of their origin.

The wine industry, like all other long-term food industries, is market or consumer driven, not production driven, and unlike most other agricultural commodities, wine markets are subject to periodic changes in demand or preferred styles. The change from white to red wine consumption in the 1970s, from Riesling (1960s), to Muscats (1970s), to Chardonnay (1980s), to Sauvignon blanc (1990s) and Pinot gris (2000s), certainly in all English-speaking markets, was not due to production planning by growers or winemakers but was caused by market demands and the changing taste in wine.

It will be noted that the group of light-bodied, aromatic white wines appears more volatile and their trends shorter-lived than other wine types, although the recent rise in consumption of Pinot noir and Italian and Spanish reds, as well as the increase in demand for softer red wines from the Rhône and southern regions of France would suggest that fashion trends affect red wines as well.

ILLUSTRATION 1.7 Winter lime-sulphur spray application, Waiheke, NZ.

ILLUSTRATION 1.8   Decomposing lava at 800m altitude. Etna, Sicily.

Growers planting new vineyards should take note of this, especially if their production expands beyond their local markets.

The system of sustainable viticulture, introduced for export wines in some countries, is a part of an effort to moderate the most negative environmental impacts of the established, industrial system of grape and wine production. It involves restrictions on the type, frequency and volume of agrochemicals and toxic sprays used in the vineyards, as well as reductions in potential environmental pollution caused by winery waste and effluent. For full registration, the grower has to keep records of all vineyard operations and sprays used and conform to a point scoring system that imposes a ban on some and places restrictions on other agrochemicals used. Export certification issued for example to New Zealand winemakers from the 2011 vintage will be conditional on achieving full registration of the local, sustainable programme. The sustainable certification system is certainly a step in the right direction, but it is not well understood in most markets and falls well short in requirements of the organic, bio-grow or bio-dynamic systems, as far as the environmental impact or carbon neutrality are concerned.

The organic and bio-grow registrations involve establishing agronomic systems that incorporate multi-cultural plantings, using crops and cover crops between vines and prohibiting all but naturally occurring elements for disease control and fertilization of vineyard soils. All man-made agrochemicals or fertilizers are replaced with preparations made from natural sources, elements like sulphur or copper, whilst fertilization relies on compost enriched with animal manure algae and the incorporation of green cover crops, grown usually amongst vines and around the vineyard perimeter. Common cover crops used by organic growers include legumes, nitrogen fixing clovers, various grains and a wide range of flowering plants and herbs (white, yellow and blue flowers preferred) to promote greater soil health and fertility and to provide habitat for a wide range of beneficial insect predators, as well as to foster greater microbial activity in the soil and bio-diversity in the vineyard environment.

In more recent years, there has been a significant increase in the organic use of the various naturally occurring and beneficial fungi, rhizomes for roots and Trichoderma species (T. viridiae) for the control of invasive, pathogenic fungi, such as botrytis, eutypa, botryspheria, armillaria and others. Added to soil or composts (rhizomes) or sprayed on the canopy, pruning wounds or delivered into trunks of vines via long-term release dowels (trichoderma), these methods of control of pathogens are registered for organic use in Europe, Australasia and some states of the USA. The organic and especially the bio-grow registrations are recognized in most countries and all key markets where wine is sold.

The bio-dynamic concept of viticulture incorporates all the key aspects of the organic/bio-grow models, coupled with Steiner School principles of balanced energy within the living environment as a whole. The principles of bio-dynamic grape growing have a foundation in a bio-dynamic calendar of moon phases for all vineyard or cellar operations, with the most appropriate periods allocated to all activities throughout the year, from pruning to harvest. The aim is to achieve a balance between the vines and all principal influences in the

ILLUSTRATION 1.9 Coastal clay loam soils with high level of salinity in Pakino region, Sicily.

ILLUSTRATION 1.10 Alluvial brown loam soil over sandstone base rich in calcium carbonate, Avola region, Sicily.

environment, such as light, air, soil, water and minerals, that nourish the vine, and bio-diversity, which improves vine's immune system and the general health and fertility of the soil.

As in the organic concept, bio-dynamic grape growing utilizes natural preparations (500, 501, algae, herb extracts etc.), which are applied to soil surface and vine canopy at the appropriate times to boost natural plant immunity or to stimulate soil micro-biology. In the case of wine, the aim is to achieve a true expression of terroir. Unlike other systems already described, the bio-dynamic regime often eliminates the use of machinery and deep soil preparations (ploughing, ripping etc.) in an effort to minimise disturbance to the established soil profile environment.

Both systems involve increased levels of hand labour input, require in-depth knowledge of the local environment and of the compatibility between vines and endemic or introduced companion plants.

The results of either the organic or the bio-dynamic approach to grape growing can be spectacular, as far as the environmental ecology and wine purity are concerned; however, it should be recognized that conversion from the industrial system to either will take a number of years to complete and may contain significant hazards along the way. Perhaps the best way to approach the conversion process is by gradual, step-by-step reductions in and replacement of agrochemicals with non-toxic organic materials or application of the organic and bio-dynamic regimes right from the start, during the establishment period of a new vineyard.

CHAPTER 2

Winter

Vine dormancy, morphology and physiology of the roots.

The onset of dormancy is indicated by cessation of shoot and leaf growth, senescence of basal leaf area, progressive lignification of canes from base the upward and the downward migration of reserve energy (carbohydrates) into the storage parts of the vine in the autumn months.

Full winter dormancy is achieved after complete leaf fall and hardening of canes, when temperatures drop below 10 degrees C (mean ambient air) or 6–8 degrees C (mean soil temperature) in the root zone. The optimum length of dormancy period for vines is stated as 60–90 days at 6–8 degrees C or less. It will be noted that cool climate region adapted varieties (Pinots, Riesling or Chardonnay) will show root growth and root metabolic activity at the lower range of root zone temperatures, whilst warm climate adapted varieties (Grenache, Mourvèdre or Primitivo) will not be active until higher soil temperatures are reached. The winter chill resistance of different varieties adapted to different temperature zones follows a similar pattern.

As indicated earlier, all growth and metabolic processes in the shoots and leaf area will cease after full dormancy is reached and all available carbohydrates were transferred into the permanent, woody parts of the vine and roots. The same cannot be said of roots, whose flush of growth occurs during these two growth periods; many of the complex root metabolic processes that occur in between continue irrespective of the reducing temperature of the surrounding soil or the air mass above the ground. It is worth noting that all energy and carbohydrates/nutrition required for root growth, water uptake and root metabolism are generated during the growing season and supplied from shoots/leaf area above ground. The morphology, physiology and the various metabolic processes of vine roots are perhaps less understood and studied than those occurring above ground, yet these metabolic processes play a key role in the vine's survival, water and nutrient uptake, carbohydrate partitioning and hormone synthesis and release—all of which influence the vine's potential to grow and produce fruit.

The dormancy itself is initiated and caused by increased concentrations of dormancy hormones (growth inhibitors) released from the seeds into the vine shoots in summer, from the hard seed stage of veraison onward. Likewise, shoot growth is promoted by the release of growth hormones in the shoot tips. Both the synthesis and the sequential release of hormones are credited with balancing the vine's potential for growth vigor (shoot growth) and fruitfulness (seed maturity), thus influencing potential yield and wine quality.

The process of carbohydrate partitioning, that is the division of fixed (stored) and available (water soluble) reserves of growth energy/nutrition, as well as the release of growth stimulants and growth inhibitors (hormones), occurs due to the vine's natural tendency to balance out the two competing functions—shoot elongation and growth of seeds. Both of these were required for vine's survival in its original forest habitats. The vine achieves its survival goal by balancing the annual amount/synthesis of each group of hormones, depending on the concentration of both during the previous season's canopy and fruit, i.e. the number of growing points and seeds. The sequential release and activation of both hormone groups, sealed during winter storage in multi-layered envelopes, occurs during the growing season due to the action of a series of enzymes, activated in their turn by increasing soil temperature in the root zone.

The principal environmental conditions that affect the rates of root growth and its varied metabolic processes include soil temperature, soil condition/biological health and physical soil structure, availability of energy/nutrition from stored reserves and the availability of water and oxygen in the root zone environment—all required for the varied processes described above. Potential problems can usually be avoided by appropriate site selection, creating an optimal soil drainage/temperature environment in the root zone by utilizing agronomic methods that promote the greatest possible diversity in soil biology and balanced nutrition, as well as by adapting viticultural practices that achieve balanced vine growth and moderate yields.

Root morphology and water/nutrient uptake.

The root system forms an essential part of the vine's anatomy, anchoring the plant and linking it with and drawing nourishment from the soil of its origin. In modern times, the rootstock also fulfills the function of protecting the European vine from destruction by phylloxera and if well chosen, it moderates the impact of water stress or excessive or insufficient vigor potential of the site. Vine roots have the capacity to penetrate deep into the soil profile and whilst the majority of the root mass will be found in the top 1–1.5 metre depth, other deeper roots, thus the overall root morphology, play an essential role in supplying a balanced and complete nutrition and water uptake, which proves especially important during the periods of heat and water stress in late summer.

Depending on the physical architecture of the soil structure, microbial soil health, vine age and the rainfall distribution, which will be region and season specific, most of the water and nutrient-rich water solutions will be taken up from the upper strata of the soil profile. With increasing vine age or under conditions of increased water demand during late summer and autumn, the vine will rely more and more on its deeper roots to gain sufficient water, nitrogen and mineral nutrients for its growth and fruit development.

Grape vines adapt well to periods of gradual increase in drought or water deficiency. This adaptation involves the flexibility of the root morphology, which efficiently exploits the soil profile and applies a control mechanism to limit excessive water loss through its leaves. The vine's drought adaptation also imposes a well defined priority order in supplying energy, nutrition and water to its various anatomical organs. As regards water usage, the shoot tip, young and mature leaf area, the tendrils and the flowers or fruit are not equal. The vine's drought adaptation, next to light sensitivity and growth polarity tendencies, is the key survival factor that guarantees its longevity and influences yield and composition of the berry and thus the potential wine quality.

Vineyard soils tend to be a complex mixture of solid particles of varying sizes, shapes, density and surface textures, reflecting the history of the vineyard's geological formation and subsequent evolution, breakdown and erosions. Together, the soil particles form the physical architecture of the soil profile in which empty spaces or soil pores are formed. The three-dimensional architecture of these spaces creates a network of connecting pores throughout the soil profile that is unique to each site. It is in these pores that the soil's reservoir of air, water and many water soluble nutrients are found, as well as the spaces through which the vine roots grow.

The pore sizes and their distribution, as well as the total volume of soil available, determine the capacity of a given soil to hold water, how tightly it is held and the degree of resistance the soil profile offers to roots. The drainage characteristics of each soil type and site, as well as the ease or difficulty with which gas exchanges between the aerial and root environment take place, has a great impact on root growth and the rates of metabolic processes within.

In principle, the more rapid the oxygen/carbon dioxide exchange rate, the greater the root growth and greater its metabolic rates. Root growth and metabolism are closely related to water and nutrient uptake, as well as the synthesis of organic compounds that are unique to roots.

To take up water and nutrients from the soil environment, the roots have to be in close contact with the soil's water solutions. Mobility of water toward roots is limited due to soil's resistance to liquid flow over larger distances, especially in a lateral direction. The root surface area determines the overall potential for water/nutrient uptake, whilst root diameter influences the efficiency of water transport within the plant itself. The actual extraction rates are controlled by the water pressure gradient (negative water pressure) created by water evaporation from the leaves, and the solubility gradients that exist between the solutions within the plant and those being taken up from the soil.

The actual distribution (morphology) of root growth is greatly influenced by gravity and by the resistance offered to roots by variable size soil pores. In reality, roots grow along the path of least resistance: compact, dense soils offer greater resistance than soils of more fragmented or friable, looser structure. It is also of interest that roots cannot grow through a rigid opening that is of smaller diameter than their own. When a growing root cannot find a suitably large pore, it will divide into smaller laterals and will continue dividing until all available spaces are filled. Unless vine roots are grown in a perfectly homogeneous soil environment, such as sand, the root system will explore and exploit the soil profile of each site in an irregular fashion that will be site specific. Its exploitative role will be limited by its ability to divide, whilst its ability to deliver water over required distances will depend on retaining sufficient diameter to overcome internal resistance, i.e., friction within its vascular system.

Root division is in part also determined by the inherited genetics of any plant species; root geotrophy (orientation) and tendency to divide show various degrees at the inter-specific relationship amongst the Vitis groups of species. For example, Galet's work in France shows this genetically imposed range of variability in root mass shape and rates of division amongst both European V. vinifera and the American native species V. riparia, V. rupestris and V. berlandieri. The practical application of his work is in the selection of appropriate rootstocks for a chosen site, especially when drought resistance or moderation of vigor is required. In other words, a given tendency in the rootstock toward a certain root shape and fineness or coarseness of its roots must be combined with the

ILLUSTRATION 2.1 Root exploration of soil profile.

appropriate structural type of soil and climatic conditions of the site. Neither of these must be allowed to conflict or override the tendency in question.

For example, the desirable deep penetrating and highly divided roots of drought-resistant rootstocks (V. berlandieri type) will be seriously impaired by excessive availability of water and soil fertility, leading to excessive growth vigor of highly water/nitrogen demanding canopies—thus a self-generating new cycle of vigor. Compare this scenario with balanced small vines, whose canopies are limited in size (1–1.5m square per 1 kg of fruit), composed of mature leaf area (60 days plus old) and grown under the progressive soil drying regime, thus under mild water stress conditions. In practice, this requires careful site selection, appropriate soil preparation and vine spacing, coupled with sound rootstock choice and long-term stability of canopy and yield size.

Not only the ratio of root surface to its weight, but also root diameter can play a role in controlling potential vine vigor. For example, a root of 0.2 mm diameter will be the limit for water uptake, whilst a 0.1 mm diameter is the limit for the uptake of nitrogen. This would suggest that water and nitrogen uptakes are different processes, yet, as practical observation shows, the two are closely linked. This observation also suggests that potential vine vigor can be moderated by either increased density of root zone soil profile or by the reduction of available soil volume, both forcing changes to root morphology by changing the soil environment.

The resulting moderate-vigor vines, if grown under a balanced regime, achieve a root morphology that favours roots of high surface to weight ratio, with a corresponding increase in drought and heat resistance, which is known to improve fruitfulness, fruit maturity and potential wine quality.

Field observation also suggests that balanced, mature and smaller vines planted at increased density simulate physiological responses usually observed in vines under mild water stress, without achieving the high rates of respiration stress observed in hot climates or when there are severe shortages of water. Such response can be highly desirable in advancing physiological fruit maturity, even in regions of higher rainfall and cooler climate or in seasons of reduced heat summations in warm climate vineyards.

The vine's physiological response, similar to water stress induced from roots, triggers diminishing shoot growth rates, changes to leaf size and reductions in respiration rates, all of which cause a quantitative shift in carbohydrate partitioning in favour of an increased number of seeds at the expense of growth vigor. The chemistry involved is well beyond the scope of this book; however, the key indicators of this process are the increased concentration and levels of abscisic acid and diminishing amounts of free nitrates and giberelic acid in the shoot tissue.

For the purposes of this book the definition of a balanced vine used throughout is

a) a mature vine of 12–15 years after planting, fully cropping and occupying all of the space allocated at planting, ranging from 1 to 3 metres square of vineyard surface per vine.

b) a vine with a ratio of 5:1 or greater of perennial to annual parts (weight to weight).
c) a vine of moderate vigor and stable ratio of 1–1.5m square of canopy for each 1 kg of fruit.
d) a vine whose root morphology is dominated by roots of high surface to weight ratio, occupying no less than 1.5–2m depth of soil profile.
e) a vine with a canopy of no less than 1 meter in height and the following composition of young (less than 60 days) to old (above 60 days) leaves during the season:

growth stage	young	old
flowering/ fruit set	75%	25%
berry growth stage	50%	50%
veraison	25%	75%
fruit maturation period	0–10%	90–100%

Individual components of the so-called balanced vine will be covered in greater practical detail in chapters that follow.

Principal pruning and canopy training systems.

The evolution of pruning techniques for grape vines have since antiquity reflected the evolution of viticulture as a whole, from its simplest form by the early semi-nomadic farmers to the present day complex systems, required for the large scale, mechanized and industrial production of grapes. The spread of grape cultivation from Asia Minor and the Mediterranean basin to cooler climates further north forced the growers to modify the pruning of vines from the stake supported, spur pruned Goblet to the cane pruned Guyot, to achieve profitable yields in areas where basal buds were less fruitful. More recently, with the advent of grafting on phylloxera resistant rootstocks and widespread mechanization of agriculture, changes to vine spacing and increased individual vine size allowed growers to reduce establishment and labour costs, thus increase the size of their vineyard plantings. The numerous pruning regimes observed today are also the result of region-specific traditions, the increasing demand for greater wine volumes and improvements to wine quality by the growing markets, as well as adapting viticulture to increasingly marginal and extreme conditions in both 'Old' and 'New' World countries.

In principle, the lower and shorter pruning systems are common in warmer, more arid regions, whilst the higher and long cane pruned systems will be found in cooler climate areas of higher rainfall. Increased planting density of 5,000 or more vines per hectare and lower per vine bud numbers, coupled with greater labour costs, are typical for quality wine regions. In contrast, the lower cost driven, high-volume production is notable for low planting density and multi-plane pruning and training systems, which often include the use of split-canopy design.

Irrespective of the actual pruning system used, the main goal is to maintain a balance between the number of buds (retained shoots) and the site/vine age-specific potential of vines to grow. Quality pruning will achieve full expansion of the vine frame into the allocated space, maintain a uniform and well distributed canopy and appropriate yields of fully ripened fruit, year after year. Experienced pruners will achieve all these goals without inflicting major pruning wounds, avoiding the future need for large remedial cuts to permanent wood.

Both the size and the precision of cuts made during the winter pruning of vines directly influences the per bud fruitfulness along the retained spurs and canes. The pruning technique and its quality as regards the size and proximity of cuts made to old, perennial wood is also reflected in the extent of internal die-back in vine spurs, cordons and trunks, as well as the likelihood of the decline of mature vines due to infections by invasive fungi, such as eutypa, botryspheria, armillaria or esca. In general principle, the closer or larger the cut made close to old wood, the greater the chance of infection and internal die-back. From the illustrations below it will be seen how extensive and damaging such die-back can be to the vascular system of vines.

In most cases seen throughout the viticultural world, the internal die-back and the action of invasive fungi is often the most common cause of the loss of productive vines rather than excessive vine age alone. Avoiding pruning cuts across the central vascular system of old wood, as often happens in reconstructive

ILLUSTRATION 2.2 Balanced VSP canopy.

pruning of misshaped vines, is important. An effort should also be made in greater use of green training in summer and early detection of infected vines (leaf symptoms), so that these vines can be cut back and re-trained before declining too far and infecting the nearby vines. The use of poorly trained, unskilled labour for winter pruning or the lack of appropriate sanitation of pruning wounds cannot be recommended in any vineyard, irrespective of vine age. A balanced canopy of a mature vine should achieve close to 1 metre in length, i.e. 12–14 leaf, with uniform diameter shoots and should produce 2 full clusters of fruit, with a low amount of sub-lateral growth. Should the grower achieve this goal by balanced winter pruning, much of the remedial summer work, such as heavy trimming, lateral, shoot/leaf or fruit removal will be avoided later on.

Young vine winter pruning requires caution, as the apparent early vigor often leads growers to retain excessive bud number and fruit in the first four or five years. Excessive buds, poorly matched with young vine's potential to expand its frame, root system and crop, leads to vine stunting and in extreme cases, to premature vine death. In practical terms, a healthy young vine can double its annual biomass in each of the first four development years. Retained bud numbers and gradual vine frame and bunch number increases should reflect this goal.

Useful advice is to increase bud numbers evenly and gradually during the early years after planting (5–7) and to keep one cluster of grapes on each fully grown shoot. Remember that the expansion of the perennial vine frame and roots requires carbohydrates and energy from stored reserves that are scarce in young vines. One simple equation of the vine's potential to grow is a the total of available stored reserves, plus the current season's photosynthetic surpluses, minus crop. Early indications of over-pruning are blind budding, short and uneven shoot length, symptoms of early water stress, poor cane lignification and delayed fruit maturity at harvest.

As a general guide, balanced pruning retains from 10–20 buds per square metre of soil surface in high density plantings, whilst mature vines grown in low density vineyards are pruned to 30–35 buds per square metre. Another way to calculate the appropriate bud numbers for mature vines is to retain 30–35 buds (low density) or 10–20 buds (high density) to each kg of pruning weight.

The given range of bud loading are of course region and variety specific. Subject to vine vigor, the region's climate and the pruning system used, the large bunch varieties such as Müller-Thurgau or Malbec are pruned to the lower bud number range, and small bunch varieties like Cabernet and Riesling, at the higher end.

Gobelet spur system with semi-dispersed or stake-supported canopy.

This is an ancient spur pruned system, best suited for high density plantings in warm/hot climate regions, where basal buds are fruitful. Bush pruning or Gobelet is common in the south of France, Spain and south Italy, introduced here by the Greeks and Romans before the advent of trellis systems being well adapted to hot, drought conditions or steep terrains.

TABLE 2.1 Bud numbers retained under different pruning systems in high and low density plantings and canopy exposure to direct light under the different canopy systems (after Jackson).

Pruning and canopy type	Vine spacing in metres.	Vines per ha.	Buds per ha.	Exposed leaf area in square metres per ha.
Bordeaux cane & VSP	1 × 1	10,000	140,000	15,000
Guyot cane & VSP	1 × 1	10,000	100,000	15,000
Gobelet spur & stake	1 × 1	10,000	80,000	9,000
Mosel cane & stake	1 × 1	10,000	140,000	12,000
Bilateral Guyot cane & VSP	2.7 × 1.5	2,500	48,000	10,500
Sylvoz bent cane & split VSP	2.7 × 1.5	2,500	66,000	13,000
Scott Henry cane & split VSP	2.7 × 1.5	2,500	66,000	13,500
Geneva double cordon/spur & canopy.	3.0 × 1.5	2,220	90,000	22,000
Quadrilateral cordon/spur & VS	P 3.0 × 1.5	2,220	50,000	9,000
Lyra bilateral cane & split VSP	3.2 × 1.5	2,080	81,000	17,000

(after D. Jackson 1998).

Relatively inexpensive to establish, the system relies less on mechanization and more on inputs by hand labour for pruning, shoot tying and harvest. A further advantage of the Gobelet system is that it achieves greater vine longevity on account of a high ratio of perenniality to annuality (carbohydrate reserves) and requires minimum injuries to older wood during winter pruning.

Pruned on a single plain of the head region, it also suppresses vine's apical dominance. Its canopy is either tied to a stake or free formed into a semi-dispersed canopy in an umbrella shape, which offers the greatest interception of light (as a percentage of surface area planted) whilst it protects the fruit from excessive exposure to sun and thus to sunburn.

In the Rhône valley of France, the 4–5 main shoots are tied to a stake, thus improving the fruitfulness of the buds for the next season. In Jerez in Spain or Sicily in Italy, the canopies are free formed into the umbrella shape, as the fruitfulness of basal buds is high.

Bordeaux bilateral cane system with or without replacement spurs and VSP canopy.

Sometimes known as the double Guyot, the Bordeaux cane pruning system is well suited for high density plantings in cool climate regions of higher seasonal rainfall, where buds in the middle of the cane (position 4–8) are the most fruitful. Increased yield per vine due to increased total bud number is an advantage,

ILLUSTRATION 2.3 Gobelet vine (before pruning).

ILLUSTRATION 2.4 Gobelet vine (after pruning).

ILLUSTRATION 2.5 Gobelet vine in summer.

ILLUSTRATION 2.6 Semi-dispersed canopy of Gobelet.

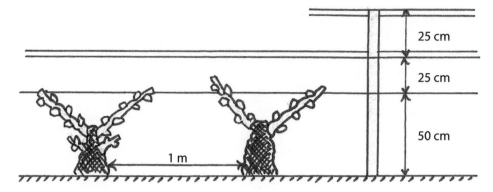

FIGURE 2.1 Bordeaux cane pruned vine with or without spurs in winter.

if compared to the traditional Guyot of Burgundy. Increased shoot density per metre of canopy and vigor/competition in vine centres and cane ends requires additional hand labour in Bordeaux, if similar canopy homogeneity to Burgundy is to be achieved.

A common problem encountered with the bilateral Bordeaux cane can be fruit overcrowding, especially with large bunch cultivars like Merlot or Malbec. Unlike the small cluster Pinot in the Guyot pruned vineyard of Burgundy, the Bordeaux cane pruning requires significant 'green harvest' fruit removal to avoid fruit overcrowding and excessive yields (limits of 35–40hl/ha in some A.C. areas in Bordeaux), where extra production per vine is thought to harm the resulting wine quality.

ILLUSTRATION 2.7 Cane-pruned vine with VSP canopy in summer, Bordeaux.

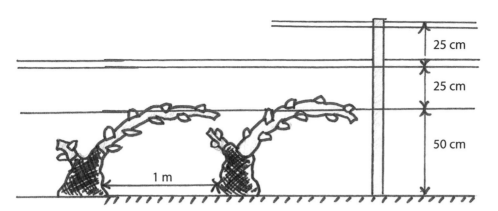

FIGURE 2.2 Guyot pruned vine in winter.

Guyot cane system with replacement spur and VSP canopy.

The system of Prof. Guyot, established in the early 19th century in Burgundy, is considered the best for all cool climate regions with high density planting on low to medium vigor sites. From a purely physiological point of view, the unilateral Guyot system achieves the best vigor distribution and the highest quality canopy environment/fruit distribution, whilst it clearly separates the replacement shoots (on the spur) from the fruit bearing shoots (on the cane).

Guyot pruned vines are usually trained on VSP canopy and like their Bordeaux counterparts, achieve high levels of exposure of leaf area to direct light. If the canopy is well managed, it also provides close to optimal levels of fruit exposure (direct and reflected) on both sides of the canopy.

ILLUSTRATION 2.8 Guyot vine with VSP canopy in summer, New Zealand.

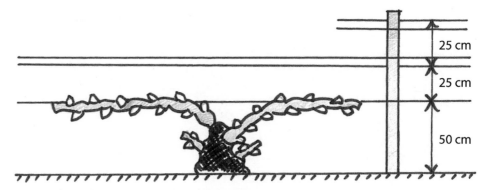

FIGURE 2.3 Double Guyot, long cane with spurs and VSP canopy training in winter.

Double Guyot, long cane pruning with spurs and VSP canopy training.

Often used in medium or lower vine density vineyards in cool climates, the bilateral cane system provides lower establishment cost, higher per vine yield and has the flexibility of a higher canopy panel, which makes possible the introduction of mechanized harvesting and the use of standard farm machinery.

Germany and northern Italy in Europe and California, South Africa, Australia or New Zealand have adopted this system since 1970s. Modification of this system with an arched cane is well suited to the climate and varieties of Alsace and Piedmonte, where the extra length of the cane is required to achieve fruitfulness for local varieties, such as Riesling, Gewürztraminer and Nebbiolo.

The long arched cane version has also the advantage of superior vigor distribution and suppression of the vine's apical dominance, if compared to the same length when using horizontal canes.

ILLUSTRATION 2.9 Double Guyot, cane system with VSP canopy in summer, New Zealand.

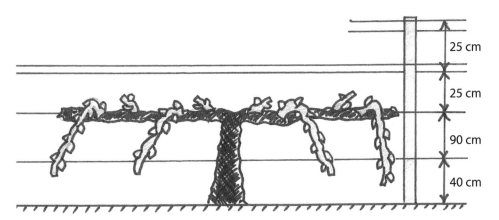

FIGURE 2.4 Bilateral cordon with 'Sylvoz', bent cane pruning and VSP canopy in winter.

Bilateral cordon with spurs or 'Sylvoz' bent canes, VSP canopy.

The combination of spurs and bent canes on a cordon system of pruning is best suited for low vine density, wide spaced vineyards in cool climate regions (Germany/Austria/northern Italy), utilizing the increased yield potential and advantages of exposure of VSP canopy training.

It can be established on medium high or high cane or cordon systems. The higher version, sometimes known as the similar 'Moser' system, is better adapted to warmer regions or varieties for red wines. Its V-shaped, umbrella-like and semi-dispersed canopy offers a well ventilated, cooler environment that protects the fruit from excessive exposure and sunburn. The cheaper and simpler 3-wire canopy support system for the canopy of the 'Moser' variation on a high cordon with spurs found favour in many vineyards of California, Australia and northeastern Italy.

ILLUSTRATION 2.10 Bilateral cordon with 'Sylvoz' system after harvest, Marche.

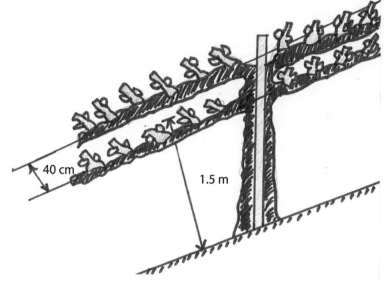

FIGURE 2.5 Geneva double curtain vine in winter.

Quadrilateral cordon system with spurs and 'Geneva' double curtain canopy in winter.

Developed in New York State by Prof. Shaulis, this system utilizes the extra space offered by wide spacing, low vine density planting and the hanging curtain of canopy for maximum interception of light. Well suited to hot climates, irrigated vineyards of high soil fertility and high volume/low cost production, this system

ILLUSTRATION 2.11 Geneva double curtain canopy system in summer, Friuli.

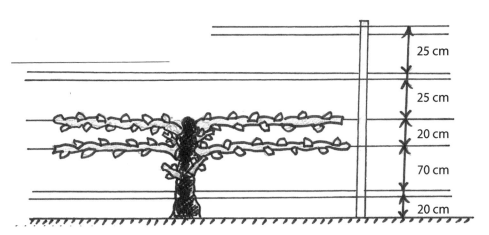

FIGURE 2.6 Scott Henry pruned vine in winter.

is relatively easy to establish or to maintain, proving popular in regions where production is high and wine prices are modest. Irrigated regions of California, Australia or Veneto and Friuli in Italy often employ the 'Geneva' curtain system for large-scale plantings.

Bilateral, cane-pruned 'Scott Henry' system with VSP split canopy.

Another modification of the multiple cordon or cane system of pruning, this system utilizes the splitting of the standard VSP canopy (half up and half down), offering double the bud/shoot number retained per linear metre of trellis, hence significant potential for yield increase and light exposure.

ILLUSTRATION 2.12 Scott Henry trained canopy in summer, New Zealand.

Designed by the Australian Dr. R. Smart in the 1970s, this cane-pruned system is best in cool climate regions, in sites with high vigor potential and the aspiration for high yields. The Scott Henry pruning system creates two tiers of canes/fruit zones with the canopy from the upper pair of canes trained in standard VSP, whilst the canopy from the lower pair of canes is forced downward and secured in this position by additional catch wires. In general, the concept works well, as long as the two canopies are well separated at flowering and maintained afterward. In practice, the system demands high labour input, whilst the lower canopy and its later maturing fruit can become suppressed with time, as the vine's apical dominance asserts itself.

Quadrilateral cane or cordon-spur pruning system with VSP or V-shape canopy in winter.

This is another of the more productive systems from the warm/hot climate regions of Australia and California, where low density, medium high fruit wire and mechanization of the vineyard operations and harvests are preferred. If dry farmed or grown on low to medium fertility sites, this system is often coupled with standard VSP training of the canopy. In more vigorous sites, regions with increased rainfall or, where high vigor rootstocks and grape varieties are planted, the canopy regime is easily modified into an open V-shape by introducing T-sections that support the canopy further apart. Such modification, often seen in vineyards of California, offers the grower an opportunity to increase bud/shoot number per metre of canopy, without causing excessive leaf to leaf shading or, excessive exposure of fruit to sun and fruit overcrowding.

FIGURE 2.7 Quadrilateral, cordon-spur pruned vine in winter.

ILLUSTRATION 2.13 Quadrilateral cane with VSP canopy in summer, New Zealand.

German 'Mosel' bent double cane, 'Gobelet' system with stake-supported, semi-dispersed canopy.

This is a modification of the ancient Gobelet bush vine system, adapted to cool climate regions by retaining long canes and stakes for shoot support, allowing high planting density and a well ventilated, illuminated canopy, including the fruit zone. It is thought that this system of vine pruning and training was introduced into the Mosel River valley in Germany by the Greeks, well ahead of the introduction of trellised vineyards along the Rhine by the Romans. This system is labour intensive but well suited to steep slopes and high quality varieties, such as Riesling.

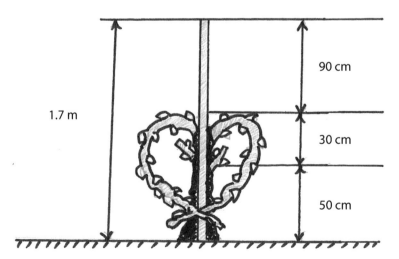

FIGURE 2.8 Mosel bent cane pruned vine in winter.

ILLUSTRATION 2.14 Mosel bent cane, stake trained vine canopy in summer, Germany.

Bilateral cane 'Lyre' trellis with inclined VSP canopy.

The Lyre system of training vines, developed by Dr. A. Carbonneau of Bordeaux University, offers a combination of wider spacing between rows and closer in-row planting, with two VSP canopy panels inclined away from the vine row centre. The vines are planted in a single row, 1 m apart and inclined in their turn

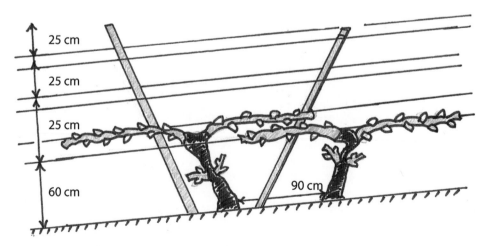

FIGURE 2.9 Bilateral Guyot cane with Lyre trellis, inclined canopy system in winter.

ILLUSTRATION 2.15 Bilateral Guyot cane with Lyre trellis and inclined canopy, France.

to the opposite sides of the 'lyre' trellis. The alternative use of this system is to plant two vine rows in staggered fashion and train the canopy in the inclined fashion, with the panels ending up opposite each other. Although expensive to establish, the Lyre system of trellis and canopy training is easier to maintain, produces more uniformly ripened fruit and is generally preferred to other split systems, such the Scott Henry or Geneva double curtain. The split canopy 'lyre' system is typically found in the New World, cool climate regions, such as Oregon, New Zealand or Tasmania. It can be difficult to mechanize, unless modified to allow for mechanical harvests.

Rootstocks for different soils and climates.

The urgent need to combat the rapid decline of French vineyards in the late decades of the 19th century, due to the devastation of own-rooted V. vinifera by the phylloxera louse (P. vastatrix), was the reason for research into and later utilisation of species of resistant American wild vines as rootstocks. The earliest research focused on inter-specific hybridization of the two species of vines, but when the resulting plants failed to produce an acceptable quality of wine, the present method of budding or grafting one onto the other was adopted. The group of American species targeted were the V. riparia, V. rupestris, V. berlandieri, as well as selections of their crossings by plant scientists, that included

Millardet, Kober, Couderc, Schwarzmann, Paulsen, Richter and Ruggeri, their names still attached to rootstocks today. Their efforts were successful beyond belief and by the time of the first International Symposium on rootstocks in Sicily (1911), the European varieties were saved and vineyard plantings in France recovered to their pre-phylloxera levels soon after. Since that time, the grafting or budding methods were adopted throughout the world and own-rooted varieties of V. vinifera are limited to outlying areas in Chile, Washington State, South Australia and southern areas of New Zealand.

Rootstocks, apart from their phylloxera resistance, provided the European vines with increased growth vigor, on account of all of the American species being more vigorous than the V. vinifera. For this reason rootstocks should be described as more or less invigorating and devigouration of grafted vines should not be expected. What follows is a list of the principal rootstocks used in more recent times, providing a guide to their origin, key tendencies and characteristics that may have an effect on the variety above. The word guide is important here, as all rootstocks will adapt to soil and climatic conditions of the soils in which they are planted and their affinity with all varieties or conditions cannot be guaranteed. Hence the guide, rather than definitive instruction for their use, is presented here. The safest solution to finding the best combinations lies in a long-term and variety-/region-specific research. Growers are well advised to research local experience if available or to establish their own trials and to research the literature that is available amongst the references and bibliography listed in Appendix 1.

ILLUSTRATION 2.16 Wild growing bush of American rootstock.

What can be said of rootstocks with greater certainty is that they will demonstrate their natural tendencies and adaptations in growth vigor, length of their vegetative cycle, drought resistance and root morphology in appropriate soils and climatic zones, all else being equal. Many will show greater ability for water or nutrient uptake, whilst others will display greater resistance to specialised soil conditions, such as the presence of active lime, salinity or low pH.

ARG-1 or AXR, Ganzin No1 (Aramon × V. rupestris)

Vigorous in growth, shows affinity with most varieties and supports high yields. Proved less resistant to phylloxera and nematodes in Europe and California, prefers deep, well moisture-supplied soils of medium to heavy density. Not well suited to dry soils or those of elevated active lime content. Not recommended on account of its susceptibility to phylloxera.

Couderc 3309 (V. riparia × V. rupestris)

Low to moderate vigor, suited to deeper and fertile soils, well supplied with moisture. Medium in vigor, grafts easily, high resistance to phylloxera, less so to nematodes or presence of active lime. Has tendency to improve set, can reach high crop levels, but shows difficulties in uptake of potassium in heavy clays and will be susceptible to drought stress in shallow, well drained soil profiles or regions of arid climate.

Fercal (V. berlandieri 333 E.M. × Colombard BC1)

Modern French rootstock with good tolerance to drought conditions and presence of active lime. Suits well drained, medium depth soil profiles and can advance maturity on account of its short vegetative cycle. Prone to magnesium deficiencies in volcanic, acid soils. Moderate to high vigor in fertile, deep soil profiles.

Gravesac (16-149 × Couderc 3309)

Another modern rootstock from Bordeaux, good tolerance to low pH soils, moderately vigorous, short vegetative cycle and so promotes early fruit maturity. Considered similar to SO-4 Oppenheim, but preferred to it in Bordeaux.

Kober 5BB (V. berlandieri × V. riparia)

Selected as one of the better rootstocks for calcareous soils and well drained sites. Vigorous in growth with moderate growth cycle, it has affinity with most varieties. Popular in cool climate regions of northern Europe, but its strong spring growth can cause problems with fruit set and in dry soils, reduced uptake of potassium also. Not suited to steep slopes, arid sites or regions of severe winter chill. In Germany it tolerates winter temperatures of minus 8 degrees C, but not greater. Good resistance to phylloxera and nematodes.

Kober 125 AA (V. berlandieri × V. riparia)

In most respects similar to the Kober 5BB but, with greater resistance to drought, tendency to higher vigor also. Well suited to poor, sloping or stony soils and higher cropping varieties, such as Sylvaner, Müller-Thurgau or Sangiovese.

Portalis or Riparia Gloire de Montpellier (V. riparia)

Superb, low vigor rootstock, well suited to deep, fertile soils that are well moisture supplied.

Its short vegetative cycle promotes early ripeness, not suited to arid soils or shallow soil profiles. Excess thickening bellow graft union is typical. It has excellent resistance to phylloxera and good affinity with most varieties.

Richter 110-R (V. berlandieri × V. rupestris)

Vigorous and resistant to drought conditions, moderate tolerance to active lime (12% or less), good resistance to phylloxera, well suited to shallow, sloping and dry soil profiles. In Germany it is recommended ahead of the other Richter selections, such as R-99, R-37 or R-44.

Rupestris St. George or Rupestris du Lot (V. rupestris)

Versatile rootstock for most varieties on well drained sites of deep profiles. High resistance to phylloxera and drought conditions, supports vigorous growth over its long vegetative cycle. Not well suited for varieties with irregular set or those grown in fertile soils in cool climates, does well with Cabernet and Zinfandel (Primitivo) in California.

Schwarzmann (V. riparia × V. rupestris)

Popular, drought resistant rootstock with affinity to sandy or calcareous soils. Moderate in vigor, shows high resistance to phylloxera, best suited to lighter soils and warm positions.

SO-4 Oppenheim (V. berlandieri × V. riparia)

Recently regaining popularity in Germany and France on account of its tolerance to a wide range of soils, short vegetative cycle and its ability to advance maturity at harvest. Moderate to high vigor, recommended for varieties with irregular set, but not for arid sites and dry soils. High resistance to phylloxera and nematodes. In Californian it is sometimes confused with 5-C Teleki.

143-A or Aripa (Aramon × V. riparia)

A low vigor rootstock with good affinity with most varieties, supports fruit set well and promotes high yields. Well suited to deep, moist and fertile soils, not suited to arid or dry conditions. Shows only moderate resistance to phylloxera and low tolerance to active lime.

5C-Teleki (V. berlandieri × V. riparia)

Like Kober 5BB, this all-round rootstock was selected for its tolerance to a wide range of soils and affinity with most varieties. Tolerant of calcareous soils with elevated active lime, its short vegetative cycle promotes early ripeness. High resistance to phylloxera and moderate vigor in growth. The new selections of 5C, such as Geisenheim 6 and 10, are increasingly popular in Germany.

26-G (Trollinger × V. riparia)

Originally from the Geisenheim Institute, it is considered the best Riesling rootstock for steep slopes and stony soils, such as those found in the Mosel region. In Austria it is often used for Veltliner, has moderate vigor and a short vegetative cycle, hence brings ripeness forward. Suited to a wide range of well drained, low pH lighter soils, has good drought resistance. High resistance to phylloxera and moderate resistance to active lime, not best suited for calcareous soils.

101-14 Millardet (V. riparia × V. rupestris)

Moderate-vigor rootstock with affinity with wide range of varieties and does well in deep, well drained soils. High resistance to phylloxera, only moderate to nematodes. Short vegetative cycle and well suited to cool climate regions, excellent affinity with Cabernet and the rest of the Bordeaux varieties due to its ability to bring harvest ripeness forward.

420-A Millardet e de Grasset (V. berlandieri × V. riparia)

Moderate to high vigor with highly explorative but less exploitative, sparse root growth. Does well in deeper, warmer and well drained soil profiles. Reduced uptake of potassium can be a problem in heavier clay soils and the length of its vegetative cycle makes it less well suited for late ripening varieties in cool climate regions.

161-49 Couderc (V. berlandieri × V. riparia)

High-vigor rootstock with a balanced nutrient uptake, best in well drained, warm and deep soil. It is drought resistant, supporting high yields in Cabernet grown on the valley floor of Napa Valley in California. Known to be sensitive to 'tillosis' in wet, clay soils in France.

140-Ruggeri (V. berlandieri × V. rupestris)

Vigorous in growth, resistant to drought and salinity in South Africa. Less well suited to cool climate regions such as Bordeaux. Known to promote irregular set with Trebbiano and Mourvèdre varieties in Italy or France and in clay soils susceptible to reduced uptake of potassium and magnesium with all varieties.

1103-Paulsen (V. berlandieri × V. rupestris)

Mostly preferred ahead of similar group of Paulsen selections, such as 775-P, 779-P or 1447-P, the 1103-P shows high vigor, aggressive root growth and a high degree of resistance to drought, salinity or high levels of active lime. Its long vegetative cycle makes it best suited to warm climate regions and earlier maturing varieties. Known to have a balanced uptake of nutrients in France and southern regions of Italy.

41-B Millardet e de Grasset (Chasselas × Berlandieri)

Moderate to high vigor in growth, best suited to well drained positions, it is drought resistant and has high tolerance of increased levels of active lime. Less well suited to wet and cold soil profiles, its long vegetative cycle is known to delay fruit maturity. 41-B is not efficient in uptake of potassium in clay soils.

Pathogenic fungi and bacteria affecting vines, pruning sanitation.

The field of the various disorders or diseases caused by a number of insects, fungi, bacteria and viruses is large enough to fill several volumes of the size of this book. Fortunately, a more specialised literature on the subject exists already and the best and most up-to-date advice on this topic is available to growers from local technical advisers, suppliers of the various treatments on the market and from the bibliography in Appendix 1. In general terms, the authors' view is that prevention tends to be better than cure and that an epidemic-scale occurrence of any pest, pathogen or physiological disorder usually has other underlying causes. Poor soil and vine health, inappropriate agricultural interventions, poor vine management, such as overcropping and depleted old wood reserves of carbohydrates or pollution of the vineyard environment with soil toxins, are all known to impact on the plant's immune system, making the vine more susceptible to secondary infections by pathogens. On the other hand, well grown and managed vines, that are not subject to injuries or severe environmental stresses and those receiving balanced nutrition, appear far more resistant to the same pathogens. A high degree of bio-diversity in the vineyards, including healthy populations of beneficial soil microbes and predator insects, make it harder for epidemics to occur.

Amongst the more commonly occurring fungal diseases affecting vines are the brown, gray or black rots that attack young leaves, fruit or both. Botrytis, Anthracnose, Aspergillus, Diploidia, Cladosporium and related species of pathogenic fungi, are usually responsible. Amongst the mildews, the 'powdery' (oidium or Uncinula necator) and 'downy' mildew (Plasmopara viticola), are commonly known. Less common, but lot more dangerous and difficult to eradicate, unless recognized and treated early, is the 'dead arm' (Phomopsis) and 'collar rot' (Phytophtora), both of which attack perennial wood and can cause vine decline if not stopped early on. 'Crown gall' is another disorder of vines, initially caused by injury to base part of the trunk, which is later infected by soil-borne bacteria, whose toxins cause abnormal swelling and distorted growth in the infected area.

The grower will also note that unnecessary large wounds inflicted during remedial or simply bad winter pruning not only cause die-back within the permanent wood, which obstructs the internal vascular system in trunks and cordon arms of a vine, but also leaves the injured vine wide open to an invasive group of fungi that are known to cause fatal decline of infected vines. It is this group of pathogenic fungi that will be the focus of our attention from here on, as they are without a doubt a major cause of vine death in all parts of the viticultural world.

The destructive power and effects of Eutypa (E. lata), Botryspheria or Italian 'eska' was known for some time to accompany the decline of many old vines. More recently, these and related species of invasive fungi were in fact identified as a primary cause of decline in apparently healthy vines of prime, productive age. The 'black goo' epidemic in California during 1990s or the long recognized 'eska' decline of young and old vines in Italy are just two examples of the direct relationship between a group of invasive fungi and vine decline.

One of the contributing factors in both cases may well have been linked to the rapid expansion of plantings in both countries, the shortage of well grown, grafted vines from commercial nurseries and perhaps, less than ideal site preparation and care for young vineyards. Surveys of nursery plants in Europe, South Africa, Italy and California show that these pathogenic fungi are present around the injured, graft area of the vine in groups of related genera, rather than as a

ILLUSTRATIONS 2.17 AND 2.18 Large wounds after poor or re-constructive winter pruning.

single species. Furthermore, despite their presence in all sampled plants, the vine decline symptoms of growth stunting and vine death, due to asphyxiation of the central vascular system effected some, but not other vines.

From experience in California we learned that aggressive forms of agriculture with ample nutrition, conservative pruning and cropping without environmental stresses is in fact capable of restoring many of the 'black goo' vines into a full growing and cropping state, whilst small percentage of the vines under the same care and conditions have declined and had to be replaced.

The 'eska' disease in Italy appears to favour some varieties more than others and mostly appears to infect vines of 10 years of age and over. So far, it appears resistant to the application of agricultural means alone, but clearly its strongest expression of symptoms and progress amongst healthy vines seems to occur in the years of greatest stress. It would seem, that avoiding large pruning wounds, severe nutritional or water deficiencies and conservative cropping regime all help to reduce symptoms but, offer no cure.

Some promising work with the introduction of the occupying Trichoderma fungi on pruning wounds, or in the form of slow-release dowels into vine trunks appears to limit the most severe symptoms of eska/eutypa on leaves, shoots or clusters, especially if infection is detected and treated early.

The work of Prof. Mugnai at Florence University or Prof. Gubler and L. Morton in field trials in California suggests that Trichoderma treatments show benefit, possibly by boosting the vine's immune system and making the spread of the pathogenic fungi slower and more difficult. However, they do not appear to eradicate these fungi at this early stage of their work. More positive results were obtained by Dr. J. Hunt in the laboratory and in his long-term field trials in New Zealand and Australia since the late 1980s.

In practical terms, one of the most popular and effective treatments for 'eska' by the growers seems to be the removal of the infected, permanent parts of the cordons, if symptoms are detected early, or to cut the trunk of the declining vine near ground level and to re-construct the vine frame from the new growth. It has been estimated that 'eska' and 'eutypa' infections penetrate some 30–40 cm

ILLUSTRATIONS 2.19 AND 2.20 Symptoms of eska decline, Tuscany (foliar and cane symptoms).

ILLUSTRATION 2.21 Sanitation with latex paint containing fungicide, Napa Valley.

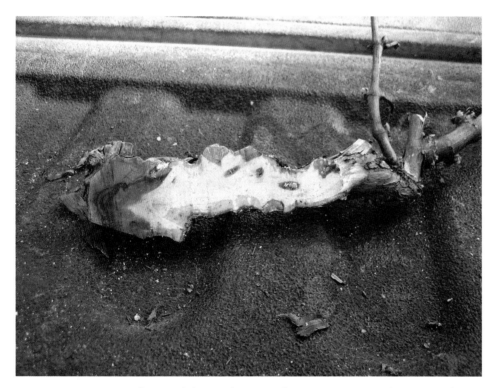

ILLUSTRATION 2.22 Internal tissue damage due to severe pruning wounds to old wood.

ILLUSTRATION 2.23 Winter pruning in Napa Valley, California.

further into perennial wood each year; thus early detection of symptoms seems most important.

The grower should also bear in mind that the infections of pathogenic fungi occur during or just after the first heavy early winter rains, known to be the optimal spore release period, and that large unprotected pruning wounds to old wood offer greater opportunity than the smaller diameter cuts to 1- or 2-year-old wood. Delaying pruning until after the major release of spores, avoiding large cuts to old wood and the application of sanitation as soon as cuts are made all help to prevent infections. The most common sanitation of pruning wounds involves either applications of a latex barrier with fungicide containing fruit tree paints or applications of a Trichoderma containing paint.

It goes without saying that dead or infected vines should be removed from the vineyard and burnt, to prevent the spread of infection into healthy vines.

CHAPTER 3

Spring

Budburst and early shoot growth.

The sequential process of budburst in the spring is triggered by rising soil temperature in the root zone and the activation of growth hormones. The upward flow of sap and early bud swell are the first indications that budburst is about to happen. The process continues through the stages of advanced bud swell and the appearance of the serrated edge of the first leaf. Until this stage of development,

ILLUSTRATION 3.1   Spring bud burst, Tuscany.

the shoot and all of its morphological parts including flowers are protected by inner bud wool and outside scales. This protection from cold weather ceases once the first leaf and the shoot tip emerge.

The active temperature for vine shoot elongation and initial leaf growth varies between varieties, lower for cool climate adapted ones, such as Riesling, and higher for hot climate varieties, such as Grenache, Nero d'Avola or Sangiovese. In broad terms, the starting point for activity is said to be in the region of 10 degrees C for shoots and close to 6–8 degrees C for root growth.

The early shoot growth is entirely dependent on a supply of carbohydrates and energy from stored reserves in the perennial wood of the vine and sufficient ambient air temperature. The maximum shoot growth rates are not achieved until the temperature reaches the high 20s, although moderate growth occurs in the 15–20 degrees C range also. Given a balanced pruning in winter, with respect to bud numbers or growing points versus the potential vine vigor, the shoot evolution shows moderate growth rates. Excessive vigor, due to insufficient growing points, causes the push from secondary buds and suckers from latent buds on the trunk or cordons. Numerous sub-laterals push along the shoot also indicate high vigor potential of both the season and the vine.

On the other hand, excessive bud numbers left in winter will translate into stunted growth and pronounced effects of polarity on the vine in the spring.

Either condition can be corrected by appropriate suckering and shoot thinning at the 3–5 leaf stage, so that more balanced growth rates are restored and vigor distribution improved. Early canopy management should focus on

TABLE 3.1 Stages of seasonal vine growth (after Baggiolini). (English translation from left to right)

00	03	09	11	13	15	19	55–57	65–69
dormancy	bud swell	budburst	separation of the first leaf	3rd leaf stage	5th leaf stage	9th leaf stage	start of flowering, cap fall	finish of flowering

73	75	77	79–80	81–83	85
fruit set, shatter	pea berry size, seed growth	berry swell, hard seed stage	bunch closure	veraison, colour stage	start of final grape maturation

ILLUSTRATION 3.2 Basal bud push, due to retaining short spurs in winter, Montalcino.

ILLUSTRATION 3.3 Poor growth due to excessive bud number left during winter pruning.

ILLUSTRATION 3.4 Premature shoot thinning, Tuscany.

ILLUSTRATION 3.5 Vines before well-timed shoot thinning, Tuscany.

ILLUSTRATION 3.6 Vines after balanced shoot thinning, Tuscany.

achieving uniformity of shoot length, directing energy to fruitful shoots and growth of flower clusters, whilst maintaining a high-quality canopy environment as far as the ventilation and light exposure are concerned.

Some of the climatic hazards during the springtime include frosts and wind damage to shoots due to rapidly changing weather patterns, especially in frost prone zones and wind exposed sites. Frosts of −1 degree C or greater will damage shoots and potential crop, whilst strong winds can break shoots of the more brittle varieties, such as Syrah or Malbec. Some form of frost protection and wind breaks should be installed in all areas affected by spring frosts and strong, prevailing winds and early canopy support should be employed.

Early canopy support at the 6–8 leaf stage is important for training of all VSP systems, to encourage maximum growth rates of shoots, for which a vertical position is the best. Trained and supported this way, the growth energy is not wasted on sub-lateral growth, and an open, well ventilated canopy is created, providing optimal conditions for flower initiation in the buds for the next season. Two pairs of foliage wires are sufficient to support and secure a vertical canopy over a length of 12–14 leaves, without allowing the shoots to bend over or to break.

In exposed areas the use of wire clips for greater shoot security is recommended. Under favorable conditions, a full height canopy should be ready for trimming by flowering and fruit set.

The desirable rate of balanced shoot growth during spring is in the range of 30–40 cm per month. The rates of growth will depend on water and nutrition availability, as well as on the weather. Supplementary irrigation combined

ILLUSTRATION 3.7 Spring frost damage.

ILLUSTRATION 3.8 Well-supported spring VSP canopy, Tuscany.

ILLUSTRATION 3.9 Spring canopy secured with wire clips, Tuscany.

with spring foliar fertilization can be useful, should the growth rates lag behind this range.

Symptoms of 'spring fever' growth distortions, chlorosis and marginal burn of leaves can appear in seasons of cold spring nights alternating with hot days, causing a physiological response in vines that find it difficult to take up and trans-locate nutrients within the rapidly expanding canopy of leaves.

Applications of seaweed preparations and foliar sprays of a complete, balanced nutrient mix, helps to moderate the most severe symptoms of this disorder, with normal growth rates and vine health returning with more settled and warmer weather. It will be noted that some varieties are more sensitive to 'spring fever' than others. Cabernet, Sangiovese or Petit verdot are, for example, more sensitive than Riesling or Chardonnay. On reaching optimal height and canopy size in late spring, balanced vines will reach trimming height some 10–12 weeks after budburst. The timing of trimming should ideally be at flowering or early fruit set, to cause a temporary flow of carbohydrates downward toward the fruit clusters.

Removal of lower sub-laterals, which compete for water and nutrients with nearby clusters, is done at this time for the same reason. From this point onward, the grower will aim to maintain the size and canopy density more or less constant, to create the optimal conditions for the growth, ventilation and maturation of the grapes without undue competition from vegetative growth. Trimming of the top and sides of the canopy is repeated as required during the post-flowering, berry growth period, whilst canopy environment is kept open at 1–2 leaf depth to maximise the effects of sunlight and disease control.

ILLUSTRATION 3.10 Foliar symptoms of 'spring fever' disorder, Sangiovese.

ILLUSTRATION 3.11 Vines trimmed at flowering/early fruit set, Tuscany.

Bud fruitfulness, flowering and fruit set.

Potential flower buds form in the axils of the leaves as the shoot develops in the months leading up to flowering. The various morphological parts of the bud can be seen and identified even in dormancy. The larger, or primary, bud formed remains dormant until the next spring, whilst the adjacent bud may grow to form a sub-lateral shoot. Within the larger or primary dormant bud, there are potentially three growing points with the central one being the most fruitful. The other two, known as secondary and tertiary buds, are significantly smaller and always less fruitful. At budburst in the spring after formation, the primary bud grows, and unless it is damaged, the other two will remain dormant. Some varieties have the natural tendency to produce multiple shoots from the three possible points of growth, the Pinot group and Traminer being good examples. This may also be a more common occurrence on excessively vigorous vines or shoots damaged by spring frosts. In this latter scenario, the grower can expect a yield reduction of up to 50–75% of average crop.

The inflorescences, or flower initiates, begin to develop in buds that grow in the spring and early summer, a full year ahead of flowering. This process, called bud differentiation, is completed within six to eight weeks, whilst the adjacent leaf is some 5–7 cm in diameter. The flower initiates begin as tendrils tipped with a group of specialised cells capable of growing into flower clusters, thus potential crop. Insufficient length of full dormancy, hormonal imbalances, low temperatures, excessive shading, high nitrate shoot content found in vigorous shoots and

ILLUSTRATION 3.12 Grape vine flower, Sangiovese.

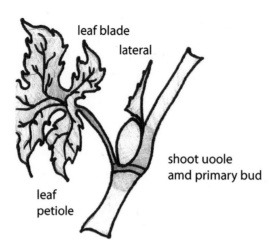

1 Primary bud
2.-3 Secondary, tertiary buds
4 Leaf petiole
5 Lateral
6 Internode

FIGURE 3.1 Primary bud, leaf, lateral and secondary buds in vines (after M. Fregoni 2001).

water or nutritional stress can all diminish the chances of bud differentiation and flower formation during the current as well as the following season.

Optimal temperatures for successful initiation of flowers is in the range of 22–25 degrees C, a sufficient supply of water or nutrients and direct light falling on the bud that is undergoing differentiation being all important. These conditions should not prove difficult to replicate in most vine growing regions during average seasons, where moderate rates of spring growth occur.

There are indications that cool climate adapted varieties are less sensitive to a lower temperature range, whilst those from hot climates require greater light intensity and heat up to 28–30 degrees C. Flower clusters can be seen on the fruitful shoots beginning in early spring, growing to full size but not opening until some 6–8 weeks after budburst. The normal rate of growth of a flower cluster relies on balanced water and nutrient supply, whilst excessive vigor is not helpful. Some varieties, for example Merlot or Chardonnay, are more sensitive than others to so-called nitrate toxicity, and reductions to nitrogen-rich spring fertilization or irrigation close to flowering may be required.

Over the rather inconspicuous grape vine flower a cap is positioned, and when the anthers are ready to release pollen the cap falls off and pollen is released onto the stigma. Cap fall indicates full bloom, and for grape vines the wind assisted pollination lasts 3–4 days under favorable, warm and dry conditions or under cooler or wet conditions up to 10 days. Flowers that fail to pollinate drop off and shatter some 2 weeks after general flowering or at best form seedless berries that remain small and often fail to ripen, as in 'coulure' and 'millerandage'. The formation of the berry and its normal growth are dependent on the presence and growth of the seed, the size of both being closely related.

The great majority of varieties depend on pollination and seeds for their fertility and berry growth. The exception are the self-fertile forms or bio-types that can form seeds without cap fall and the so-called seedless hybrid varieties that are usually grown as table grapes. In general practice, depending on spring climate, a well trained canopy of quality environment and growth at the moderate pace of 30–40 cm per month provides a supportive environment for a good set of some 100 berries per cluster and a harvest bunch weight of 100–200 g each, variety specific.

As mentioned previously, trimming at flowering/set and the removal of competing sub-laterals in the fruit zone can improve set, especially under adverse climatic conditions. As with trimming, a temporary interruption of top growth can also be achieved by using growth retardant or CCC (chlormequat) where permitted or by girdling of trunks or canes, which interrupts the outflow of photosynthates from the green canopy into the trunk and roots.

Another possible cause of yield reduction during the fruit set and during the berry growth stage is a physiological disorder called 'stem necrosis', which affects the individual bunch o r berry stems. Recent German studies suggest that it is caused by imbalances in the ratio of available potassium to magnesium in the rapidly expanding tissues of flower stems just before or during flowering. The drying out of parts of stems, sometimes accompanied by secondary infection by botrytis, appears to be in reality caused by stem damage earlier on. The 'early stem necrosis' (ESN) or 'late stem necrosis' (LSN) appear to be one and the same disorder occurring at different points of time during cluster evolution, at flowering/fruit set or as late as after veraison.

Stem necrosis affects some varieties more than others, and rootstocks usually play part as well. The recommendation is to minimise the chance of occurrence and to maintain an optimal ratio of the two nutrients at 5 K:1 Mg. The ratio is more important than the absolute volume. Moreover, tissue analyses confirm the highest demand for magnesium before flowering and for potassium after fruit set to veraison. In practical terms, 2–3 foliar applications of magnesium sulphate (1% concentration) and 2–3 applications of foliar potassium sulphate (same concentration), usually applied with sulphur, can be effective in preventing this problem on most sites. It is worth noting that these two elements are antagonistic (chemically incompatible) and combined application onto soil or under irrigation emitters appears less effective than applications on leaf. It is also worth noting that organic forms of both nutrients are available for organic or biodynamic vineyards. Indicator varieties include Riesling and Cabernet sauvignon.

From tissue analyses made during different growth phases of vines, the observation can be made that the nutritional requirements of grape vines are not constant during the growing season, that the uptake of water and nutrients is in fact demand driven and changes, depending on the vine's growth stage and metabolic activity at the time. The macro- and micro-nutrient status in the vine tissue thus changes to reflect the plant's needs within the limitations of availability of the nutrient required. It is also true that not all sites and region-specific climatic conditions or rootstocks are equal in ability to take up, store and release balanced nutrition as required by the vines. Early identification of imbalances in the plant tissue, growth rates or leaf, and other symptoms indicating nutritional deficiencies in a given site or variety, is of great value in providing the grower with site-/variety-/season-specific experience.

The late flowering or second set clusters often develop in seasons of reduced primary set in the upper parts of the canopy, causing a problem, especially if machine harvesting is used. Some varieties have a greater tendency to produce second set (Pinot, Traminer, Riesling etc.). Given its smaller size, position and late ripening, second set fruit is easily rejected during hand harvests. Second set fruit is less common in seasons of well set primary clusters.

ILLUSTRATION 3.13 Combination of cluster/tendril indicating low bud fruitfulness.

TABLE 3.2 Example of foliar analyses report from Italy (after A. Paoletti).

Analysis	Result	Threshold	Interpretation	Comment
Nitrogen %	0.46	0.50	lower average	Slight chlorosis likely, foliar addition after fruit set recommended.
Phosphorus %	0.09	0.15	low	Foliar symptoms likely, apply on leaf after fruit set.
Potassium %	1.12	1.0	low/average	Appears sufficient in absolute amount, given the ratio with magnesium, leaf application after fruit set is recommended.
Sulphur %	0.21	0.20	normal	No application required.
Calcium %	2.98	1.20	high	In combination with high magnesium and the likely high active lime level in the soil it explains the low uptake of potassium. Avoid future additions.
Magnesium %	1.94	0.50	high	Sufficient in absolute amount, but the ratio to potassium is high (2:1). Confirms that foliar K is needed.
Boron (ppm)	32.3	30	normal	No applications are required.
Iron (ppm)	31	90 +	very low	Likely to cause chlorosis on leaf, autumn soil application is suggested.
Manganese (ppm)	196	25	high	Typical of soils with high level of salinity, could also be caused by previous treatments of canopy.
Zinc (ppm)	58.8	30	high/normal	No treatment required.
Copper (ppm)	135	6	high	Previous foliar treatments is the likely cause.
K:Mg ratio	0.6	4:1	low ratio of K to Mg	Likely to cause symptoms of K deficiency on leaf, fruit phenolic maturity and reduced water/heat stress efficiency of the vines. Foliar applications of K from fruit set to veraison required to restore balance.
N:K ratio	0.4		normal	Balanced at low levels of nutrients.
N:P ratio	5.1		normal	Balanced at low levels of nutrients.
P:Zn ratio	15		normal	Balanced at low/moderate levels.

ILLUSTRATION 3.14 Poor set due to coloure disorder, Merlot.

Canopy construction, canopy quality and optimal size.

As indicated earlier, early spring shoot growth depends entirely on nutrition and energy from stored reserves. Carbohydrates, water and all available nutrients are mobilized by temperature-driven enzymatic action from storage in permanent parts of the vine and transported upward into growing shoots and the rapidly expanding leaf area above. Photosynthetic activity can be detected in young leaf area already, although the amount produced is insufficient to support respiration and leaf growth until leaves reach full size and a more mature age.

The morphology of the shoots is visible from the first month after bud burst. An elongating shoot tip, young leaf area, tendrils and clusters all unfold and grow in a regular pattern along the shoot. With increasing temperature and light intensity, the shoot growth accelerates, as seen by increasing inter-nodal length, until the maximum growth rates per day are reached by late spring and early summer. With increasing maturity of 40 days plus, the basal leaves begin to export surplus metabolites into actively growing points, shoot tips, fruit and young, growing leaves.

The growth of flower stems will be observed also; however, because the grape flowers originate from modified tendrils, they are not high in order of priority for the allocation of water and nutrition within the shoot, and only the surplus not used by shoot tips and young leaf area is available for them. For this reason, severe water or nutritional stresses during the rapid canopy growth before flowering will have a strongly negative effect on flower development and eventual yield.

ILLUSTRATION 3.15 High-quality spring growth in Sangiovese.

Not all spring shoots that grow are useful for canopy construction or benefit the quality of the canopy environment. The common practice is to remove shoots arising from latent buds along the trunk or cordons (suckering) and sister shoots from basal buds along the canes or spurs. The optimal time is at the 2–3 leaf stage, to focus all the growth energy into fruit-bearing, primary shoots. It will be noted that shoot elongation is greatest when the shoot canopy is upright and supported in VSP (vertical positioned canopy) by 2 or more pairs of foliage wires. Similar support in Gobelet is offered by tying the central shoots to a stake, thus avoiding wind damage and preventing excessive sub-lateral growth in the fruit zone. Here, the full bud exposure in the axils of the leaves is essential for the initiation of potential flowers for the next season. Late spring to early summer is the most crucial period for achieving high levels of bud fruitfulness for the next year's crop. Both vertical and semi-dispersed canopies of the various pruning systems shown benefit from the improved light exposure and ventilation offered by an early supported canopy. Benefits include improved bud fruitfulness and suppression of the incidence of pathogenic fungi (botrytis, oidium or peronospora), in addition to micro-climatic advantages within the optimum canopy environment achieved.

With the rapidly increasing rate of growth of shoots care must be given to canopy support during the spring and the canopy environment quality from early summer onward. Under reasonable climatic conditions and care, the shoot canopy should reach full height of 1 metre or 12–14 leaves by flowering. This gives a reasonable target of 30 cm plus shoot growth per month, which is achievable with ease, providing that the sub-lateral and sucker growth competition is prevented. The optimal canopy density is close to 10–12 shoots per metre of canopy and of 1–3 leaf depth, thus achieving close to 60% direct light leaf exposure and uniformity of shoot length. Depending on the site, variety and seasonal conditions, some shoot or lateral thinning, irrigation and foliar nutrition may be required to reach the optimal parameters described above.

Unlike the canopies of mature vines, those of young or widely spaced vines can benefit from retention of some or all sub-laterals above the fruiting zone, to achieve a continuous panel of foliage of uniform porosity that reaches the

ILLUSTRATIONS 3.16 AND 3.17 Well supported VSP canopy of high quality canopy environment, Sangiovese.

optimum of 1.5 metre square of canopy to each 1 kg of fruit produced. In mature vines or those planted in a high density of 5,000 or more per hectare, lateral removal may be required just before flowering (basal part) and just after flowering/trimming in the upper part of the canopy. Trimming the tops and sides of fully grown canopy will provide additional benefits to fruit set, if carried out at flowering and fruit set period, by reducing competition between the vegetative vine growth and fruit berry growth.

It is worth noting that a partial or full basal leaf removal in the fruit zone can have both positive and negative effects, as far as fruit set/early seed/berry growth and disease pressure are concerned. Whilst sanitation and the composition of tannins can be enhanced by light exposure of the fruit from berry set to harvest, early leaf stripping at flowering and berry set, before full seed growth is reached approximately halfway between flowering and the hard seed stage of veraison, negatively impacts seed formation and early seed growth, causing variable berry size or 'hen and chicken' symptoms. Recent studies into overexposure of fruit to direct sunlight by excessive or early leaf removal suggest that it leads in many regions of high light intensity (Italy, Spain, California and Australasia) to crop loss, sunburn damage to fruit and reduced phenol maturity of the grapes at harvest. The most noticeable effects of 'global warming' include a significant increase in the frequency of unseasonable, extreme weather conditions (dry, hot, cold and wet), increased light intensity (heat) and changes to the light spectrum (quality composition) measured in different parts of the globe.

ILLUSTRATION 3.18 Overexposed grapes after excessive leaf removal in the fruit zone, Merlot. (Note: some authors claim that early post-flowering exposure reduces the risk of sunburn and benefits berry composition at harvest, others suggest the opposite is the case.)

As in most vineyard operations, the timing is usually more important than the procedure itself and results will be region/site, variety and season dependent. In general, the canopy size should remain relatively stable from fruit set to harvest. From a metabolic point of view, the canopy should be dominated by the mature leaf area (60 days plus) from fruit set onward, providing the vines with greater heat and drought resistance, as well as much reduced water and nitrogen uptake. It will be noted that fruit set and early seed/berry growth suffer greatly if forced to compete with vigorous shoot growth or a rapidly expanding young leaf area, which is inevitably accompanied by increased free nitrate levels in the shoot, all of which reduce potential yield and delay physiological fruit maturity at harvest. Early water stress, excessive respiration rates and nutritional deficiencies all lead to reduced rates of phenolic metabolism, whilst rapid accumulation of sugar and fruit accelerates dehydration.

On the other side of this equation is the scenario where increasingly greater average leaf age and progressive soil drying, which limits the water/nitrogen uptake, create a balance between the carbohydrate and phenolic metabolic rates and thus contribute greatly to the balanced composition of earlier ripening fruit. Where required and available, appropriate irrigation regimes that replicate the natural pattern of progressive soil drying from veraison onward can be helpful in hot, dry summers and arid regions. Experience will show that berry size during its growth stage will be in direct proportion to seed size and seed number, all else being equal. Limiting factors include shortages of water and nutrition, climatic extremes, excessive lateral growth close to fruit or early removal of mature leaf area adjacent and just above the clusters at flowering or early berry growth.

ILLUSTRATION 3.19 Symptoms of severe water stress, Merlot.

TABLE 3.3 Bud fruitfulness along the cane (after Jackson & Schuster 2001).

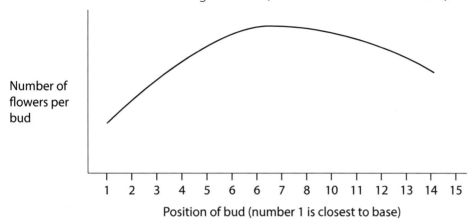

Berry set generally coincides with the switchover from the release of stored carbohydrate reserves from the vine's old wood to the utilisation of the current season's photosynthates from the leaf canopy. Further shoot, leaf or fruit growth and maturation will from now on be dependent on photosynthetic surpluses, generated by the current season's canopy of leaves.

CHAPTER 4

Summer

Berry and seed growth from fruit set to veraison.

Berry growth from fruit set to harvest maturity occurs in three stages. The first is the period of rapid growth, followed by a slower period from full seed size (hard seed cover) to the color change stage of veraison. The third and last period is one of water uptake and volume and weight gain during the berry and seed maturation period, from the end of veraison (full color) to harvest.

As both shoot and berry growth are controlled by growth regulating hormones during the initial, first growth period after set, the hormones released from the seeds are vital to the continuation of berry growth and the retention of berries on the clusters. Some varieties have the capacity to develop and grow full-size berries without seeds, presumably by genetic mutation in their

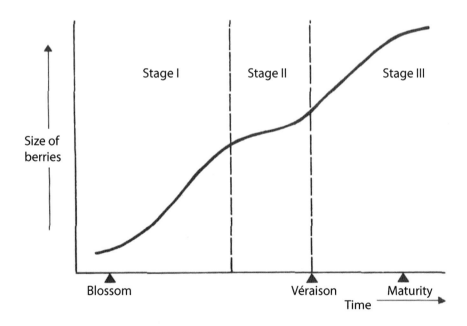

TABLE 4.1 Berry growth and weight from fruit set to veraison (after Jackson & Schuster 2001).

developmental physiology, but the great majority of wine producing grapes cannot, and thus the presence of seeds is essential for their berry survival.

During the first two growth stages to veraison, green berries grow by cell division, and due to the presence of chlorophyll and stomatal openings, they are capable of photosynthetic activity and respiration. It is also true that the berry growth after set relies mainly on nutrition from the canopy of leaves above, its own metabolic processes being insufficient by themselves.

The second growth period of the berry is slower and shorter in duration, occurring just before the end of veraison; however, the seeds develop more rapidly to their hard seed cover stage of maturity, which signals the onset of veraison, at which point the berry loses chlorophyll and its ability to grow by cell division. From here on, the berry experiences rapid inflow of water, it softens, its sugar levels begin to rise and acids to drop, and its by now translucent skin begins to accumulate pigments that give it its red or yellow color.

At the end of veraison itself, the berry acid content is at its highest, but due to dilution and the conversion of acids into insoluble salts, the overall acidity begins to decline. First there is a drop in tartaric acid to levels that remain stable later on, during the maturation period. Malic acid, on the other hand, diminishes at a constant rate by the action of enzymes, activated by the temperature of the berry. Enzymatic activity increases in periods of hot weather and slows down when temperatures decrease. For this reason the grapes grown in cool climate regions tend to retain higher levels of total acidity and higher malic acid content, whilst those from the hot climate regions experience the opposite. Simply stated, it is clear that the composition of acids and the total acidity in the berry are linked to heat summation during the post-veraison maturation period, whilst the sugar and phenolic content reflects the climatic conditions of a much longer period, from fruit set to harvest.

Respiration rates, water and nitrogen usage by vines.

Most viticultural regions of the world have an annual rainfall of 700–800 mm, which, given the appropriate site preparation and rootstock, should be sufficient to support moderate vigor and yield with most varieties cropped at 1–2 kg per vine level.

The most important factor is the seasonal distribution of the rain, with the optimal scenario being 60–70% of the rain falling in the winter and spring months, with the remaining amount through the summer and reducing greatly in the last weeks prior to harvest. Apart from the sanitation required for reduction of disease pressure, progressive soil drying in late summer and autumn will promote earlier and more complete physiological maturity in the grapes.

In hot climate regions of insufficient rainfall and a high volume of grape production, some form of supplementary irrigation will be required. Various forms of irrigation are used in different parts of the world, such as flood irrigation in Chile, overhead sprinklers or under vine drip systems in California, South

Africa and Australasia. In appellation controlled regions of the EU such as A.C. in France or D.O.C.G in Italy, irrigation is not permitted but can be used for lesser table wines. For dry farming, the best drought adapted strategies include the use of drought resistant rootstocks (V. berlandieri type), clean cultivation or closely mowed green cover, a higher density of smaller vines of low to moderate yields and semi-dispersed canopy systems, such as Gobelet. Preventing growth of shoots and laterals late in the season forms an essential part of the above regime.

Irrigation, where necessary, should be designed to match natural vine growth patterns and achieve progressive soil drying either by using a pattern of diminishing volume, or even better, by decreasing the frequency of the same volume of irrigation sets, which usually start after fruit set, during the berry growth stage and stopping them altogether around veraison, certainly before the final 6 weeks prior to harvest. Where the soil profile and summer heat permit, deep irrigation at increasingly greater intervals or alternating the side of vines per irrigation set (using two emitters per vine) will prove more effective in avoiding water stress and promoting wine quality. Frequent irrigations of smaller water volumes, under similar soil profile conditions where drainage is concerned, achieve the opposite result. Establishing a clear pattern of irrigation sets from fruit set in mid-summer as outlined above, will promote the onset of the vine's natural drought responses, irrespective of water availability later on and without the vines showing severe water stress. In short, plants like the grape vine respond better to changes in patterns of water availability, as with increasing or decreasing daylight, than to single events.

The drought responses of vines can be monitored by observing signs of the plant's conservation of available water, changes to its metabolic processes, drying up of tendrils, lignification of canes, drying of shoot tips, cessation of growth and gradual moderation of photosynthetic rates. All of these responses are beneficial to greater heat resistance and recovery after heat events, as indicated by recovery from high levels of negative water pressure in leaves (early morning versus midday variations). The resulting increase in rates of phenolic metabolism and improved rates of trans-location of metabolites from leaves to fruit, all benefit the earlier maturity of the fruit.

The actual uptake of water and nitrogen from the soil by roots is regulated by the osmotic pressure that is generated by water transpiration from the leaf canopy. Vigorously growing vines with a large canopy and a high percentage of young leaf area generate, under the hot and windy conditions of mid-summer, the greatest rates of transpiration, which often exceeds the ability of roots to absorb and deliver sufficient volume to replace the water being lost. Under these conditions (wilting point), the stomata closure stops all gas/water exchanges between the leaves and atmosphere, which leads to immediate interruption of metabolic processes and trans-location of metabolites within the vine. A useful field indicator of these shut-down events is the loss of leaf turgidity and a rapid rise in leaf temperature above the ambient air temperature. In extreme cases of repeated water or temperature stress, loss of shoot tips, basal leaves and fruit shriveling will be seen.

ILLUSTRATION 4.1 Grape bunch collapse due to failure of the berry reflux mechanism.

The gradual water loss from the berry during the last weeks before harvest should not be confused with a physiological condition causing a rapid bunch collapse near harvest, sometimes called 'water berry' or 'folletage'. Complex changes to shoot chemistry and the physical failure of the berry water uptake/loss reflux mechanism, following alternating periods of rapid water uptake and water loss during the periods of extreme heat and heavy rains, are responsible here. Merlot and Pinot noir are useful indicator varieties, whilst Cabernet and other thick-skinned, small berry varieties are less affected.

This and other damaging scenarios, involving irregular water and nitrogen uptake, are mostly encountered on sites where poor drainage and the water retentive profile of soil, coupled with heavy irrigation and/or applications of nitrogen-rich fertilizers, cause waterlogging of the root zone, combined with excess growth vigor above the ground. In practical terms, the worst case scenario described here can be moderated, or prevented altogether, by improvements to perimeter and subsoil drainage, by ripping alternate centres of the rows or by the use of lower vigor rootstocks, establishing competitive green cover and avoiding late season irrigation/fertilization of vines. It will be noted that every vineyard site has limitations, most of which can be prevented during the establishment period, or moderated by adapting site-specific agronomic and viticultural practices or by the use of rootstocks and varieties better suited to local conditions.

ILLUSTRATION 4.2 High altitude vines of Karazakis variety near ancient Troy, Turkey.

ILLUSTRATION 4.3 Coastal vineyards of ancient Thracia, Asia Minor or modern Turkey.

Photosynthesis, carbohydrates, crop levels and phenols.

Photosynthesis is the process by which all plants synthesise simple carbohydrates like sugars from water, carbon dioxide and energy from sunlight. All green, chlorophyll containing parts of the vine, leaves, shoots or unripe berries participate in this process, which provides metabolites and energy for vine growth. Efficiency of the photosynthetic process depends on the amount and quality of sunlight available, temperature, water uptake, gas exchanges between the vine and atmosphere (carbon dioxide in and oxygen out) and the outflow of photosynthates from the leaf itself. Photosynthates are produced within the leaf canopy in the form of simple sugars, fructose and glucose, which combine into sucrose and are transported into areas of greatest need or hormone pull, such as the tips of shoots, young leaves and toward seeds in the ripening fruit. Any surpluses are delivered into perennial parts of the vine, roots and trunks, and stored there in the form of starch. These so-called reserves of carbohydrates are essential for the vine's longevity, root growth, long-term fruitfulness and the next season's early canopy growth. Stored reserves are also required for root metabolism during dormancy. Not all of the stored carbohydrates are released or available each year. The actual available amount has been estimated at 20–25% of the total amount stored.

For its balanced, long-term growth and the annual maturation of fruit, the vine relies on photosynthetic surpluses from its current leaf canopy. These surpluses are generated in fully expanded, mature leaves (40 plus days old), that are sufficiently exposed to direct or reflected sunlight. Younger leaves, on account of their respiration and their own nutritional requirements act as net importers of carbohydrates, for which their modest photosynthetic production proves insufficient. With increasing leaf maturity, the production of surpluses gradually shifts upward on the shoots. It has been shown that the production from 6 mature leaves is required to support either a rapidly growing shoot with its 5–6 immature leaves or one full size cluster to maturity. The optimal shoot length of 12–14 leaves and the recommendation that post-veraison shoot growths be moderated or eliminated are based on these findings.

The rates of photosynthesis progressively decline in old, basal leaves (90 days or older), very much in line with their declining respiration rates and consumption of nutrients. This decline does not mean that such leaves are without benefit to the plant or the grower, as their production of chemical precursors of phenolic compounds (phenolic metabolism) continues to contribute to the eventual composition of the fruit, despite their declining photosynthetic rates. For improved sanitation and light exposure of ripening fruit, the grower usually removes the oldest basal leaves in the fruit zone, especially those that are trapped inside the VSP canopy or between clusters of grapes. Their loss is compensated for by the improved exposure and thus efficiency of the retained leaf area.

Fruit removal (green harvest) is one of the most effective means to maximise potential vine quality/berry composition and to achieve a balance between the canopy size and yield in any given season. It will be found that crop removal at the stage of hard seed cover is most effective in berry compositional improvements and is without the risk of compensating growth of retained fruit. This compensation effect is most pronounced in seasons of larger or more numerous seeds and less so in years of less favorable vintages. Removing crop or flowers too early risks compensation in the remaining grapes, and too late (after veraison) results in much reduced berry compositional gains. Early trials in Bordeaux suggest that the removal of flowers rather than clusters of growing berries results in reduced risk of berry size compensation. If proven long term, this method may be of most benefit for the large cluster varieties such as Merlot, Malbec or Sangiovese, where delayed fruit removal can cause significant soft fruit damage, close to veraison.

The timing of canopy/fruit manipulations by growers is of great importance, as they should aim to accommodate the natural vine tendencies of shoot/cluster growth and the corresponding requirements for its various metabolic processes, during the evolution of the growing season.

For example, late suckering and shoot or crop thinning result in uneven or stunted canopy growth, whilst late trimming of overgrown canopies diverts nutrition from flowering, fruit set and early berry growth. Poorly managed canopies often display excessive leaf to leaf shading that leads to poor fruit composition, delayed ripeness and reduction of fruitfulness in the following season.

It is worth noting that both leaf to fruit and leaf to leaf shading within the canopy produce similar results in fruit composition and both are known reduce potential wine quality.

The carbohydrate and phenolic metabolisms in the shoots are not mutually exclusive processes. However, it will be found that leaf age or climatic conditions and shoot chemistry that strongly favour one lead to reduction the in rate of the

ILLUSTRATION 4.4 Poorly managed canopy in mid-summer, Montalcino.

other. It is known that concentrations of different hormones, nitrate compounds, amino acids and temperature-driven enzymatic activity all play a key role in either promoting or retarding growth. The rates of photosynthetic activity, and the rates of assimilation of primary compounds required for synthesis of many of the phenolic compounds later found in berries, respond accordingly.

From the grower's practical point of view, it is important to recognise that rapidly expanding canopies dominated by young leaf area will promote increased respiration and photosynthetic rates, accompanied by high free nitrate content in the shoots. Smaller canopies, composed of mature leaf area, grown at a slower pace and under greater diurnal temperature variation from veraison onward, will promote the synthesis of greater amounts of aromatic, flavour carrying phenolic compounds in the resulting wine.

From the climate's point of view, it can be said that there will be a direct correlation between the total heat from flowering to harvest and the total phenolic accumulation at harvest, the sugar levels will respond most to heat between veraison and harvest, whilst total acidity and its composition (tartaric/malic,

ILLUSTRATION 4.5   Green harvest in a Sangiovese vineyard.

etc.), will be linked most closely with heat and diurnal (day/night) temperature variations during the final stages of fruit maturity.

The manipulations of canopy size and canopy quality, including average leaf age and density, are one of the key tools available to the growers. Another, is the adjustment of crop size, with both aiming to regulate vine vigor, balance shoot/leaf metabolism and thus the composition of the grapes at harvest. Early suckering, shoot thinning at 2–3 open leaf stage, rapid and well supported canopy expansion to full length at flowering, lateral removal and first trimming at flowering/fruit set, followed by green harvest of excessive or overcrowded fruit at hard seed, basal leaf removal in the fruit zone and ongoing maintenance of canopy density/size to harvest provide the kind of regime that achieves the best results.

Berry weight, volume and composition changes from veraison to harvest.

The berry volume and its weight continues to increase beyond veraison. The cell division within the berry has stopped; however, due to the influx of water and sugars and the assimilation of a complex range of phenolic compounds contributing to color, aroma and flavour, the berry volume and its weight increases throughout the maturation period to the point of full physiological maturity of the seed. Recent studies have shown that the post-veraison berry weight increase can be calculated with a high degree of accuracy by establishing berry weight at veraison (hard seed stage) and calculating the predicted weight gain of the same berry at full physiological maturity, using a variety-specific multiplication factor. It is also of interest that the two time periods of fruit set to veraison and veraison to full physiological maturity are equal in length. Because both of these observations relate to the maturity of the seed, rather than to the more climate dependent berry composition alone, they offer more dependable estimates not only of the yield, but of the precise period of time during which the full physiological maturity of the fruit will be reached in any given year.

It will also be noted that physiological maturity and the actual harvest may not coincide, as the first relates purely to final seed maturity, whilst the second is often influenced by a range of climate dependent berry composition parameters, such as sugar, acid and pH or the desired flavour profile and wine style considerations imposed by the winemaker. From the grower's point of view, the physiological fruit maturity, the fruit's physical condition and the harvest weather should carry more weight; unfortunately, often they do not. More on this and related topics in chapter V.

The final stages of berry maturation are accompanied by complex changes in the berry's composition, changes that are greatly influential on potential wine quality. Whilst the sugar and water influx into the berry is most rapid during the first weeks after veraison, the assimilation of the many phenolic compounds in the shoots and their trans-location into the fruit and final synthesis in the berry is a much slower process. Accumulation of color pigments, aroma and flavour components carries on at a relatively even pace until the harvest. The rate of the trans-location process of all the varied components later found in the berry, is influenced by climate-driven factors of temperature and water availability. Photosynthetic activity, the respiration process and concentrations of the solution within the berry itself are also important. The temperature and incidence of rain play an important part in accelerating or slowing down the berry maturation process.

On the more practical side, the canopy size and maintenance of a high-quality canopy environment, including the leaf and fruit exposure, are essential for best results. Excessive leaf to leaf or leaf to fruit shading, significant dilution due to late season rains or irrigation, as well as late season vegetative growth, will prove detrimental to berry composition and the potential wine quality.

Vigor control, irrigation and dry farming.

Some of the key elements of vigor control, dealing with the roots and canopy management were already covered in detail in earlier chapters. As an overview, the grower will observe that their soil management, fertilization regime, overall canopy size and leaf age composition, as well as their ability to distribute vine vigor during the season well, will all impact on the growth vigor in both the short and long term. With sufficient site- and variety-specific experience, the grower will be able to avoid over- or under-pruning their vines in winter and to moderate yields in abundant vintages.

ILLUSTRATION 4.6   Dry farmed Gobelet vines in ancient Thracia, Turkey.

By observation, grape growers will also note the spring growth rates of their vines and use shoot thinning, lateral and leaf removal and trimming in an effort to achieve and maintain optimal canopy size and a high-quality canopy environment. In a balanced, mature vine the vegetative growth will occur in the first half of the season, whilst after fruit set and certainly after veraison, the focus will switch to supporting the growth and maturation of the grapes.

The use of appropriate rootstock that is compatible with the climate and soils of the site and has a strong affinity with the variety grown will make this effort much easier. When available and permitted, irrigation can play an essential role in controlling vigor, supporting economically viable yields and improving potential wine quality. In arid regions of hot climate and insufficient rainfall, the flood, drip or overhead sprinkler irrigation systems prove to be valuable tools for growers.

In regions of limited or seasonal water availability and where terrain permits, the flood irrigation method of wetting the deep soil profile proves most effective in the periods of early shoot growth in the spring or during rapid berry growth after fruit set in mid-summer.

In areas of poor soil fertility and low vigor, deep, post-harvest irrigation is of most benefit. Examples of well managed flood irrigations will be found in Chile, South Africa and Australia. Southern parts of Italy and Spain offer the best possibilities for utilizing this effective and least expensive irrigation system available. Labour will be required for establishing and maintaining the flood water channels and reservoirs for water storage, but the rest carries no major cost. Utilisation of this or other systems of irrigation will of course require changes to EU regulations governing water usage in the vineyards in Europe.

In frost prone and arid regions of grape production in cool climates, the overhead sprinkler system is often used for the dual purpose of preventing frost damage to grape vines in the spring and of moderating water stress in dry summers. As with all irrigations, the sprinkler system can also deliver nutrition for the vines and prevent excessive heat stress, although care must be taken to avoid

ILLUSTRATIONS 4.7 AND 4.8 Mechanical leaf removal (before and after), Friuli.

sunburn, if sprinklers are used during the day. A sprinkler system is the most expensive type of irrigation to install, although consideration of substantial losses of fruit and shoots due to spring frosts must be included in such calculation.

The drip irrigation method also carries a high initial cost but offers greater precision in water delivery and is easier to use and maintain. It utilizes a delivery system of pipes below and above ground and one or two emitters by each vine. Another advantage is that it can be automated to deliver a precise quantity of water, over a pre-set period of time. Except in vineyards of uniform soil porosity and consistency of profile, the drip system will create a site-specific wetting pattern that can and usually will impact on both root system distribution and morphology.

An improvement to the original single emitter and single line delivery system of drip irrigation, originating from Australia, was to place two emitters and two independent water lines on both sides of the vine, thereby delivering water to alternate sides of the root system each time. Using this method, the grower is able to achieve the so-called progressive soil drying regime and replicate the effects of the natural diminishing of water availability in hot summer months. It has been shown that the vine's drought responses, which occur naturally in dry farmed vineyards, can thus be triggered by a series of irrigation sets, without the risk of severe water stress.

Irrespective of which irrigation system is used, the grower should aim to manage the vine's water stress levels during the shoot and berry rapid growth stage and moderate the effects of extreme heat, without causing late season vine growth or disruption to the natural pattern of fruit maturation after veraison. Various means of field monitoring of water loss (soil) or water usage by vines are available, including soil tension meters that indicate the availability of soil moisture in the root zone or instruments (pressure bomb) that accurately measure water usage by the vines by displaying of negative water pressure and recovery rates, measured in the vine leaves in the morning and mid-day. The grower will also be aware that site-, yield- and variety-specific experience as to volume and timing of irrigation will prove of greatest value.

The dry farming regime is the most common system used in the traditional European regions of wine production, like Italy or Spain and in some parts of the 'New World', where wine quality is of paramount importance and the climate is favorable to this system of grape growing.

Careful site selection, site preparation and the use of appropriate rootstocks, as well as favorable rain distribution during the season, are all important factors for dry farming of vines.

Soil water conservation methods, usually involving clean cultivation or regular cover crop upkeep, choice of appropriate vine spacing and vine training system and rigid control of canopy size and yield are typical in all quality dry farmed vineyards.

In the hot climates of southern France, Italy or Spain, the Gobelet or bush vine system of pruning, grafting vines on drought resistant rootstocks and planting varieties of grape vines that are well adapted to hot and dry conditions in summer proved to be the best approach. A further advantage of Gobelet pruning

ILLUSTRATION 4.9 Vineyards on the coastal hills of ancient Lydia in modern Turkey.

is that the small leaf canopy is semi-dispersed and protects fruit from direct sun during the hot summer months.

In the cooler climates of northern Italy, France or Germany and in vineyards of medium to high density, variations on the Guyot cane system pruning work well also. The combination of sufficient rainfall, well drained soils, a smaller leaf canopy and yield control are common here, whilst wine quality is second to none. Various regional adaptations will be found throughout Europe, accommodating specific requirements of the sites planted, varieties grown, traditions and the wine styles made.

CHAPTER 5

Autumn

Yield estimates and grape harvest maturity indexes.

For practical and economic reasons, all grape growers endeavour to adopt one of the available crop estimation procedures. The most obvious and simplest is to weigh grapes on delivery to the cellar at harvest, establish the net weight of each variety and charge the winemaker accordingly. Useful as it is, this method proves of little use in guiding the grower to improve their viticulture or to better understand the performance of individual sites, different pruning and vine training regimes or to establish an accurate per vine yield in their vineyard.

From the wine quality point of view, the per vine yield and its variations from vine to vine are of greater importance than averages calculated by hectare and from a large population of vines that contributed to the load being delivered. For example, if the optimum crop per vine is say 2 kg per vine, then two vines cropping at 3 kg and 1 kg respectively will not reach the standard required; in fact both will bring the potential wine quality down, despite the fact that their average is the desired 2 kg. In this scenario, the over-cropped vine's grapes will cause dilution of overall character and the grapes from the under-cropped vine will be overripe and are likely to have poor balance and be equally less desirable in aroma/flavour profile. In short, a random blending of grapes of significantly different ripeness levels in the vineyard does not produce fine wine.

Experience will demonstrate that the total yield harvested divided by the number of hectares or vines grown has little value unless the grower can ensure that all plants bear the same crop weight within a reasonable variation of 10% of the mean. The only practical way to achieve this goal is to utilise a more precise method of yield estimate or per vine yield. It should be said that for growers who wish to practice crop removal at the 'green harvest' stage near veraison, the data from the actual harvest come far too late to be of real use.

The most common, more accurate, but also more labour intensive method in present use is to design a grid of sites and variety-specific sample plots of

8–10 vines per plot. By harvesting one average size, fully formed cluster from each vine and by counting the total number of such clusters per each plot of 8–10 vines (repeated throughout all the plots per variety and site containing no less than 3 such plots), the grower will be in a position to establish the average weight of the sampled clusters. The average cluster weight is then multiplied by the total cluster number on an average plant in all plots within chosen site, to provide grower with an accurate overall yield estimate, as well as a record of per vine yield and its likely variation from vine to vine.

This method works well and has a better than 10% accuracy, but it requires a relatively large sample of grapes and gives meaningful results only if carried out close to full fruit maturity, near harvest. As in the previous case, it is of little use for 'green harvest' of excess fruit at veraison, which occurs some 2–3 months earlier. The grape wastage is usually prevented by combining the yield sampling with maturity analyses of the fruit composition in the cellar, where grapes can be utilized as a starter culture or, if the harvest is already underway, the resulting grape must is simply added into full vats already there. It is worth noting that crop reductions close to harvest, unless of major proportions, will achieve minimal changes to the composition of the remaining fruit. As a guide, a 50% reduction near harvest, brings compositional benefits similar to those a 10% crop thinning would have if done during early veraison.

The most recently developed and far less crop destructive or labour intensive method, is based on our present understanding of the relationship in the seed and berry growth physiology and how this relationship reflects on harvest berry weight. The most important benefit to grower and winemaker is that this method provides accurate data of per vine bunch harvest weight at the hard seed cover stage of veraison, some 8–12 weeks prior to actual harvest, which also happens to be the optimal period for accurate crop reduction, if required. Like the previous method, it utilizes a grid of sampling plots that are site and variety specific. Unlike the first, the procedure requires the weighing of only 2–3 clusters per plot or, 6–9 clusters per site of 1 hectare size.

From long-term studies of the physiology of berry and seed growth, we know that actual berry growth by cell division ceases at veraison and the only changes that can occur after this point in time to berry volume and weight are due to the influx of water and the trans-location of sugars. We have also noted that there is a direct relationship between the seed and berry weights at veraison and the predictable berry weight increase at full physiological maturity of the seed near harvest.

Furthermore, the length of time from fruit set to veraison corresponds exactly to the time required by the seed to reach full development or the ability to germinate. Due to the near perfect insulation of the seed from climatic influences outside of the berry, its maturation during the post-veraison period proceeds at an even and more predictable pace, if compared to that of the temperature-driven berry composition itself.

Observation of the basic phenology of the site and its variety-specific nature, such as the timing of the fruit set to hard seed cover stage of early veraison, offers the grower an opportunity to calculate the arrival of full physiological seed maturity, near which the actual harvest should take place. Both the coefficient of berry weight increase from veraison to seed maturity and the greatest potential aroma/flavour and phenolic content of the berry can thus be accurately predicted in any given season. The hard seed cover stage is easily recognized in field, by simply slicing a few berries of the chosen variety in the chosen site with a razor blade and noting the difference in resistance offered by berry flesh and the seed cover. Because all vines of the same variety grown on the same site will go through this stage of seed development within a 48-hour period of each other, the correct time period required for this calculation of yield and seed maturity should be easily identified.

At this point, the grower collects and weighs the average clusters from each sampling plot and site (large, medium and small), and when their weight is multiplied by the appropriate coefficient number, specific to each group of varieties shown in table 3, the harvest per vine yield can be established. If crop reduction is required, the per vine cluster number can be reduced to predetermined levels ('green harvest'), always making sure that the retained clusters are of similar size and are uniformly distributed through the fruit zone, without overcrowding of retained fruit.

From experience, the best instruction to workers carrying out the fruit thinning is to retain a given number of clusters of average size per vine, making sure that they are well separated and undamaged. Using this method, only the vines with surplus crop are targeted, thus bringing the average crop weight per vine closer together. It will be also observed that counting and weighing of

ILLUSTRATION 5.1 Evaluation of grapes for 'green harvest', Cabernet sauvignon.

ILLUSTRATION 5.2 Overcrowded grapes without 'green harvest,' Sangiovese.

ILLUSTRATION 5.3 Overcrowded, leaf stripped grapes with botrytis, Merlot.

ILLUSTRATION 5.4 Uniform cluster size and distribution after 'green harvest', Syrah.

under-sized fruit has no impact on the accuracy of predicting the final harvest yield, as over-sized bunches and losses due to natural causes or the selection process during harvest will compensate. On the other hand, the undersized and oversized or badly positioned and damaged fruit can also be targeted for removal during the 'green harvest' process.

The actual calculation of cluster weight:

$$\text{cluster weight at hard seed} \times \text{coefficient of weight increase} = \text{cluster weight at harvest}$$

The calculation of timing of physiological seed maturity :

$$\text{length of time from fruit set to veraison} \times 2 = \text{physiological maturity of the seed}$$

Note: The above multipliers are based on studies of vines grown under moderate, rather than extreme climatic conditions, during the final maturation period. Heavy rains, severe drought or heat, frosts and disease pressure can cause significant losses of fruit at harvest.

The data is also based on trials in vineyards of mature, healthy vines of balanced vigor, cultivated in both dry farmed and irrigated vineyards.

It is of interest to note that the period of physiological seed or berry maturity does not always coincide with the actual harvest. The determination of optimal harvest maturity by the growers and winemakers in different regions and for different varieties and wine styles made, is a complex and often subjective matter

TABLE 5.1 The varietal groups and multiplier used for each in estimating cluster weight at harvest.

Groups with examples of varieties	Multiplication coefficient of est. weight increase from veraison to harvest
Group A—small cluster, round berry, aromatic red & white (Cab. sauv., Ries., P. noir, Sauv. b., Muscat b., Chardonnay)	× 1.6–1.8
Group B—large cluster, round berry, aromatic red & white (P. verdot, Semillon, Vermentino, Cab. franc, Chenin b., Grenache)	× 2.0–2.2
Group C—small cluster, oval berry, red & white (Chasselas, Syrah, Viognier, Pugnitello)	× 1.8–2.0
Group D—large cluster, oval berry, red & white (Merlot, Malbec, Veltliner, Trebbiano, Sangiovese)	× 2.2–2.4

that involves a number of variables. The evolution of each season is unique, and even the most experienced growers will confess that they would find it difficult to recall two vintages that were exactly the same, irrespective how many harvests from the same vineyard they may remember.

The complex interaction of each site, seasonal weather and variety grown give each season its unique character and personality. Most of the growers also recognise that the moment of the actual harvest is one of the most influential of all the factors, defining the resulting wine's potential for quality. Harvest too early and sacrifice the complexity of fully mature grapes, or pick too late and lose the balance, finesse and aroma/flavour concentration and true terroir expression that give fine wine its unique personality. Growing healthy, fine grapes matured to perfection thus combines science, art and luck, coupled with a healthy dose of common sense and above all, site-specific experience.

Some of the better known grape maturation indexes, such as that of Prof. Fregoni in Italy or Prof. Glories in France, deal with the complex balances of season-specific berry constituents, such as sugar, acids and pH, as well as the many aromatic and flavour carrying phenolic compounds and their potential to reach season-specific concentrations that are extractable at harvest. As such, they prove most useful for monitoring grape maturity progress and help in choosing the best time to harvest. Apart from the weather conditions at harvest, winemakers aware of the demands imposed by wine consumers play an increasingly important role in many vineyard decisions, such as which varieties to grow, required yields and when to harvest for a specific style of wine. On the more practical level, the trend toward the super-ripe, high alcohol and densely flavoured red wines, appears to be changing, favouring the more elegant, balanced and fresher styles that offer earlier drinking appeal for wine consumers. In respect to this change, it appears likely that the physiological maturity of the grapes and the actual harvests will continue to come closer together, rather than extending the recent past trend of being weeks apart.

The direct relationship between the seed size and maturity relative to berry growth and potential harvest weight has already been detailed in previous chapters. The relationship between seed maturity and berry composition, in respect to the total extractable amount of its components, is also important when choosing the optimal harvest date. Under normal climatic conditions of an average season and balanced crop, the optimal harvest period, based on seed maturity, should be reached within 100–110 days after fruit set, irrespective of climatic zone or the grape variety grown. A study of long-term data from France, Germany, Italy or California supports this time frame, as far as the extractable phenol compounds, including anthocyanins, are concerned.

It should also be noted that the loss of the physical berry condition due to imperfect berry hydrology (dehydration under failing water reflux mechanism) in hot climates or the loss of aroma and flavour components in cool climate regions occurs due to over-maturation of fruit.

Recent work by Prof. R. Bolton at The University in California Davis suggests that significant changes to the composition of extractable organic acids occur in overripe fruit, due partly to the uptake of potassium from the mature basal leaf area (senescence) and in part due to chemical changes within the berry itself. Long-term experience, supported by the available data, points to the fact that the desirable aroma and flavour profile, as well as the extractable phenol compounds and acids, do not continue to increase with significant delays of harvest, despite possible the increase in absolute amounts present.

The above-mentioned phenol maturity index of Prof. Glories recognizes that the berry phenol content changes during the maturation period from veraison to full ripeness, in both the total and the extractable amounts of their potential maximum levels.

The potential maximum of berry tannins is largely determined by the grape variety and the seasonal climatic conditions, especially heat and rainfall. The potential total phenol content usually coincides with full technological maturity (as measured by sugar/total acid/pH), whilst phenol flavour/color maturity, as well as maximum extractable tannins or color, occur close to full physiological ripeness (seed maturity), indicated by the degradation of the grape skin's inner cell walls. Both skin and seed tannins and their extractable amounts diminish with over-ripening of grapes beyond the point of full, physiological maturity.

It should be noted be noted that the grape seeds, stems and berry skins contain different groups of tannins. Those of the stems and seeds are the more simply

TABLE 5.2 The long-term average periods of maturity in number of days from flowering to harvest for red grapes in Bordeaux and Burgundy (after Blouin & Guimberteaux, 2000).

Decade	1940s	1950s	1960s	1970s	1980s	1990s	minimum	maximum	long-term average
Bordeaux	N/A	119	112	109	110	113	104	125	113 days
Burgundy	107	106	108	103	99	100	95	122	104 days

structured pro-anthocyanidins, whilst grape skins are rich in the more complex molecules formed by tannins and sugars, simple tannins and anthocyanins or polymerized tannins with anthocyanins. The complexity of the phenol maturation does make it difficult to understand the exact nature of extractable tannins in the harvested grapes, when choosing the most appropriate vinification and extraction method.

The Glories phenol maturity index, introduced in 1990, aims to identify the total content of the different phenol compounds and combine it with their potential extractability during vinification, using extreme conditions of low pH (pH 1 and 3.2) for maximum color extraction from mashed grape skins.

Table 5.3 below contains the most important values as:

A-1	=	amount of potential anthocyanins in mg/litre at pH of 1.0 value
A-3.2	=	amount of potential anthocyanins in mg/litre at pH of 3.2 value
EA	=	extractable tannins expressed as % of the total measured
Mp	=	contribution of the seed tannins expressed as % of the total measured
Skin to juice ratio	=	indicating the level of dilution of grape juice

Prof. Fregoni's maturation index, is based is based on the relationship of the final berry composition and heat summation on the final 40 days prior to

TABLE 5.3 The maturation index of potential phenol content in grapes (after Glories)

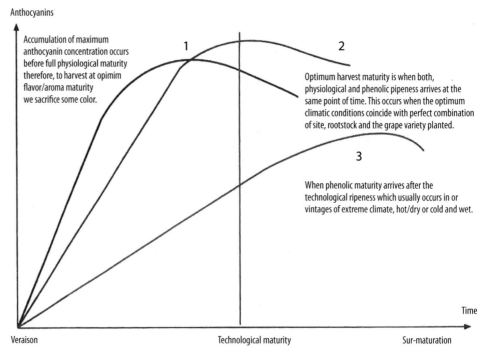

harvest. His work in Italy suggests that this period of time is the most influential in determining sugar/acid/pH/flavour balance of grapes during vintage.

In contrast to the European indexes of fruit maturity, the Australian work of Gladstone suggests a close correlation between the warmest month of the growing season and the final composition of grapes at harvest and potential wine quality. Here, as with the Fregoni index, sugar/acid/pH are the main criteria for full commercial maturity of the fruit.

Mechanical or hand harvest.

The grower's decision to use either machine or hand harvest is based on a number of practical and wine quality considerations. The factors favouring mechanical harvesting include shortages of labour in regions of large production, frequency of adverse harvest climatic conditions, such as excessive heat, cold or rain, dangers of disease pressure that may destroy the crop unless it is harvested rapidly and not least of all, the effect on the cost of the grapes being harvested.

Growers who favour hand harvests do so because the size of their vineyard and availability of labour is not an issue and because they require hand selection of grapes to achieve the wine quality level or style that justifies the higher price for their grapes. Producers of fine wines, red or white, as well as makers of specialist styles, such as Champagne or late harvest wines, insist on receiving whole clusters of undamaged fruit, which have been hand selected in the vineyard to remove the less ripe and imperfect fruit. Despite of all the recent advances in machine harvesting, these criteria can often be met by hand harvest only. It should also be noted that many of the traditional planting and pruning/training systems in use, are not designed for or best suited for machine harvests anyway.

From the wine quality point of view, the grower is required to deliver fruit of uniform ripeness, harvested at the point of optimal maturity and in the best possible physical condition. There is no doubt that significant improvements have been made in the field of mechanical harvesting since it's introduction in the 1970s. Key recent developments include much improved control over the severity of the impact of harvesters which have resulted in the reduction of damage to fruit, canopy or trellis, night harvesting to prevent overheating of fruit, more thorough removal of trash, such as leaves and introduction of optical technology for sorting out undersized, unripe fruit and second set. The introduction of field, mobile destemmer/crusher units for white grapes and temperature and atmosphere-controlled transport for white and red grapes, go a long way toward eliminating some of the key problems experienced in the past decades.

The growers and winemakers themselves can make further improvements by eliminating inferior fruit in the field, before the arrival of the harvester machines, coupled with hand sorting of delivered fruit on arrival at the cellar. The search for improvements in the technology for machine harvests is an ongoing process; however, the mechanical way of picking grapes remains largely in the domain of basic table wines, perhaps better suited for white than for red grapes. From the economic standpoint it could also be argued that the labour cost reductions are more than offset by the capital cost of buying, operating and maintaining the harvest machine itself and this, coupled with energy costs, carbon footprint and the potential damage to vines and trellis in the vineyards, can amount to considerable expense. To maximise the efficiency of machine harvests, growers must first adapt their training systems and trellis in the vineyards, so that mechanical harvesting can be applied safely.

The factors contributing to potential wine quality by hand harvesting include the use of well instructed and supervised labour in an effort to achieve precision in fruit selection during harvest and to prevent damage to grapes during field collection and transport of grapes to the cellar. The speed of harvest, the physical integrity of the fruit and fruit temperature are all important quality factors. As with machine harvests, the best winemakers also include a second sorting process in the cellar.

> "Winemakers have spent the last 100 years devising means to prevent damage to their grapes, all in the name of wine quality. Therefore, I see no sense in using harvesting machines, designed to destroy the same wine quality in seconds."
>
> (Prof. E. Peynaud of Bordeaux University).

Post-harvest vine management.

The post-harvest period under green leaf canopy provides the vine with an opportunity to redirect its seasonal, photosynthetic surpluses downward, from its canopy and lignified canes into its reserve storage areas. Depending on

ILLUSTRATION 5.5 Vines after mechanical harvest, Friuli.

climatic conditions and ambient temperatures, this process lasts some weeks during autumn and early winter, supporting the flush of root growth and boosting overall carbohydrate reserves.

For this reason, especially in low to moderate vigor sites or young vineyards, the winter pruning should be delayed until full dormancy sets in in mid-winter. With respect to pruning wounds and sanitation, autumn/early winter pruning contributes significantly to increased pressure from the invasive group of fungi (eutypa, botryspheria, armillaria etc.), whose greatest sporadication period follows the early, major winter rains and declines from then on.

It has been shown that where available a deep post-harvest irrigation is of greatest benefit in invigorating vines in dry regions or low vigor sites. The grower is well advised to observe the canopy prior to winter pruning, as poor lignification and visible drying of canes, delayed fruit maturity, as well as post-harvest canopy growth, are the usual indicators of poor canopy/yield management in the previous season. Adjustments to retained bud numbers in winter pruning, future yield levels and attention to improved vine nutrition may well be required, to restore vine balance and to improve future yield and fruit quality. The post-harvest period, with vines under green foliage, offers an opportunity to apply balanced foliar nutrition, should the problems with low yields, vigor and carbohydrate reserves persist.

Pruning weights are sometimes used as a rough guide to optimal pruning and bud number levels required. In general, a range of 20 buds (high density) and up to 30–35 buds (low density) are recommended per each kg of prunings for mature, fully established and cropping vines. Another simple way, is to observe how many full, metre-long shoots have grown and cropped the previous season, discount short shoots and retain the same bud number to that of full size shoots found.

The length of full dormancy required before initiation of pruning will be climate (rain), region, site and variety dependent however, as a general guide it is recommended that pruning not be started until after the shortest day of the year, bearing in mind that pruning and tying of vines should be well completed before bud-swell in the early spring. In bio-dynamic viticulture a format of the most favorable calendar periods is employed. In a standard, balanced vineyard the flower and fruit designated calendar days are best for pruning, whilst young or low vigor vines are best pruned on leaf or root days.

On the larger industrial scale or where the availability of skilled labour is a problem, both the mode and the timing of pruning is adjusted accordingly. For example, mechanical pre-pruning finished by hand, is possible in cordon-spur pruned vineyards, to save time and reduce the cost. A number of fully mechanized pruning methods have been tried also, usually failing on account of vine damage and poor canopy or fruit quality. The so-called 'minimal pruning' system from Australia, where green hedging replaced winter pruning altogether, was tried in the mid-80s with some promising results for 2–3 seasons after conversion. Multiple problems with sanitation, quality of interior wood structure and declining yields proved difficult to overcome, and the idea has since been abandoned.

ILLUSTRATION 5.6 'Moser' adaptation of the high Sylvoz cordon with V-canopy in Napa Valley, California.

ILLUSTRATION 5.7 Mechanical pre-pruning of cordon-spur pruned vines, Tuscany.

ILLUSTRATION 5.8 Use of electric vineyard vehicles during hand harvest. Bolgheri, Tuscany.

ILLUSTRATION 5.9 Mechanical harvest in Friuli.

ILLUSTRATION 5.10 Transfer of grapes into gondola, Friuli.

ILLUSTRATION 5.11 Guyot pruned Merlot vine, before pruning, Bolgheri.

ILLUSTRATION 5.12 Uniform bunch size and fruit ripeness of Sangiovese at harvest after earlier 'green harvest' (hard seed) Tuscany.

ILLUSTRATION 5.13 Well managed Sangiovese VSP canopy, Scansano in Maremma.

It will be observed that the different parts of the vine's morphology, perennial or annual, perform their specialised tasks but also function as a whole. All of their principal expressions, exploitative, explorative and metabolic, are synchronised in a sequential manner of growth, reproduction and dormancy in accord with genetic tendencies contained in the vine's DNA.

The complex interrelationship of the vine's perennial and annual organs can perhaps be best understood as inverted mirror images of each other (not unlike the coded script of Leonardo de Vinci), separate yet interdependent, both in the ground and above.

The roots explore and exploit the available soil profile driven by geotrophy (gravity), absorbing all water-soluble nutrients and minerals required by the organs above ground. Their efficiency and rates of extraction limited only by the physical conditions of the soil profile and by water and nutrient availability and are governed by osmotic pressure generated by the loss of water from the canopy above. The nutrition and energy for root growth come from stored carbohydrates and from those manufactured in the leaf canopy during the current season.

The most notable functions of root metabolism are the synthesis and distribution of growth promoting and growth retarding hormones, which control shoot growth vigor, fruitfulness and dormancy in the vines. Another is the carbohydrate division or 'chemical memory' of the vines from one season to the next.

In contrast to roots, the annual growth of shoots, leaves, flowers and tendrils, which explore and try to exploit the space above ground, is driven by apical dominance, phototrophy and the ability to utilise sunlight energy (photosynthesis) and carbon dioxide, whilst using water supplied by the roots to manufacture simple sugars (carbohydrates). Shoot, leaf, tendril and green berry growth is largely nitrate driven, in contrast to the carbohydrate driven growth of the roots or maturing fruit after veraison. Young leaves show greater rates of respiration and transpiration (gas exchange and water loss), whilst the mature leaf area is less efficient in photosynthetic activity (after 90 days); however, its contribution to phenol metabolism, and thus to the flavour, aroma and color composition of the ripening grapes, continues at a more even pace that is more dependent on physical conditions within the canopy than on leaf age alone. Light and temperature loading, the availability of water and shoot chemistry are the most important factors here.

The overall canopy size (relative to crop) and average leaf age thus play a crucial role during the post-veraison, maturation period of fruit and berry composition at harvest. Excessive (or insufficient) canopy size, especially if accompanied by late season growth of shoots, laterals and young leaves, results in grapes high in potential alcohol but low in phenol content. Poor canopy quality environment or size or leaf age composition lead to reduced resistance to heat or water stress, often resulting in fruit dehydration and in extreme cases, a risk of sudden bunch collapse.

A stable canopy size of optimal environment and progressively increasing leaf age from fruit set to harvest achieves the required balance between

ILLUSTRATION 5.14 Cordon spur with elongation cane, Cabernet in Napa Valley.

ILLUSTRATION 5.15 Clemens spray recovery sprayer.

carbohydrate and phenol metabolism in other words, full physiological fruit maturity, some 100–110 days after fruit set in all but extreme seasons.

If required, the crop adjustments made in the 'green harvest' also have a major impact on the eventual berry composition and the potential quality of the resulting wine. The common practice is to remove excess bunches at the early

ILLUSTRATION 5.16 Scientific study of vine metabolic processes, Bolgheri-Italy.

stage of veraison (hard seed) and to finish by full color stage. Earlier bunch thinning can result in compensation in the growth of the remaining berries, whilst removal of fruit closer to the harvest requires a greater amount of grape removal to achieve the desired compositional improvement in the remaining crop.

The natural tendencies, adaptations and region-specific mutations of grape vines provide growers with a useful field of observations and study, offering a guide to modifications of agronomic and viticultural practices, as well as allowing field selection of superior vines, best suited to each unique terroir and wine being made.

APPENDIX 1

Bibliography, Literature, References and Further Reading

Asney, W.W. (1974) Oregon climates exhibiting adaptation potential for V. vinifera. Am. J. Vit. Vol. 25/4.

Becker, N. (1985) Site selection for viticulture. Int. Symp. Cool Clim. Vit., Oregon.

Blouin J., and G. Guimberteau (2000) Maturation et maturité des raisins. Ed. Feret, Bordeaux, France.

Boselli, M. (1991) Int. Symp. Pavia 1987. Ed. Logos Int., Pavia, Italy.

Busby, J. (1833) Journal of a tour of Spain and France. Govt. Printer, Sydney, Australia.

Carbonneau, A. (1979) Research on criteria of training systems for grape vines. Plantes 29:173–185.

Carbonneau, A. (2003) Ecophysiologie de la vigne et terroir., Terroir, Zonazione-Viticultura, Ed. L'informatore Agrario, Verona, Italy.

Champagnol, F. (1984) Eléments de physiologie de la vigne et de vit. générale. Montpellier, France.

Chaptal, J.A.C., A.A. Parmentier and L. d'Ussieux (1801) Traité théorique et pratique sur la culture de la vignes. Paris, France.

Chung, C. (1983) PhD thesis, "Root Physiology." Lincoln Univ., New Zealand.

Coombes, B., and P. Dry (1992) Viticulture, Vol. II, Winetitles. Adelaide, Australia.

Crepsy, A. (2006) Manuel pratique de taille de la vigne. Oenoplurimedia, Ch. De Chaintre, France.

Dal, F., and E. Bricaud (2008) B.I.V.C. et Sicavac. Ed. Paquereau, Sancerre, Loire, France.

Fregoni, M. (1985) Viticultura generale. Reda, Roma, Italy.

Fregoni, M. (1991) Origines de la vigne et de la viticulture, Musumeci Editore, Saint-Christophe, Italy.

Fregoni, M. (1998) Viticultura di qualita. Tipographia Lama, Piacenza, Italy.

Fregoni, M., D. Schuster, and A. Paoletti (2003) Terroir, Zonazione-Viticultura. Ed. L'informatore Agrario, Verona, Italy.

Galet, P. (1979) Practical ampelography. Cornell Univ., Ithica, NY and London.

Galet, P. (2000) Precis de viticulture. Ed. JF Print., Saint-Jean de Vedas, France.

Gladstone, J. (1992) Viticulture and environment. Winetitles Adelaide, Australia.

Hameed, M.A. (1988) PhD thesis, Lincoln Univ., New Zealand.

Henri, C.C. (1993) Masters thesis, Lincoln Univ., New Zealand.

Huglin, P. (1986) Biologie et ecologie de la vigne. Ed. Payot, Lousanne, Switzerland.

Jackson, D.I. (1998) Pruning and training of vines. Vit. I, Lincoln Univ., New Zealand.

Jackson, D., and D. Schuster (2001) Production of grapes and wine in cool climates. Butterworths, Wellington, New Zealand.

Koblet, W., and I. Perret (1980) The role of old vine wood on yield and grape quality. Symp. UC Davis, California, 164–169.

Levadoux, L. (1956) Les populations sauvages et cultivées de V. vinifera. Institut nationale de la recherche agronomique. Vol. I, 59–117.

Logothetis, B. (1962) Les vignes sauvages en antique material primitive viticulture en Grece. Thessalonika, Greece, 1–43.

Meredith, C.P. (2000) Vine tendencies and grape vine genetics. UC Davis Press, California.

Morrison, J. (1990) Effect of canopy shading on grape composition. UC Davis Press, California.

Morton, L.T. (1985) Guide to viticulture east of the Rockies. Cornell Univ. Press, Ithica, NY.

May, P., and A.J. Antcliff (1964) Fruit bud initiation. Aust. Jour. Agr. Scie. 30.2:106–112.

Millardet, A. (1885) Histoire des principales variétés et espéces de vigne. Ed. Masson, Paris, France.

Pongracz, D.P. (1978) Practical viticulture. Ed. D. Philip, Cape Town, South Africa.

Poupon, P., and L. Jaquelin (1960) Vignes et vins du France. Flamarion, Paris, France.

Proessler, H. (1974) The first traces of viticulture on the Rhine. Alg. deut. Weinfach Zeit., Neustadt, Germany.

Ribereau-Gayon, J., and E. Peynaud (1961) Traité d'oenologie. Béranger, Paris, France.

Ribereau-Gayon, J., and E. Peynaud (1971) Traité d'ampelogie. Dunod, Paris, France.

Riou, C. (1994) La détermination climatique de la maturation du raisin. Commissione Europ., Paris, France.

Rowe, R. (2003) Terroir. Zonazione-Viticultura. Ed. L'informatore Agrario, Verona, Italy, 473–478.

Saurish, R.M. (1954) Dormancy in woody plants. J. of plant physiol. 5:183–204.

Schneyder, J. (1869) Culture of vines and other fruit trees in Roman times. Manière-Login, Dijon, France.

Smart, R.E. (1973) Sunlight interception in vineyards. J. of Am. Vit. Enol. 24:141–147.

Smart, R.E. (1987) Influence of sunlight on composition and quality of grapes. Acta Hort. 206:37–47.

Smart, R.E., and M. Robinson (1991) Sunlight into wine. Winetitles, Adelaide, Australia.

Trotskii, P. (1929) The phylloxera problem in Europe. J. of Entom. 15/1:1–197.

University of Bordeaux III (2002) Les territoires de la vigne et du vin. Ed. Feret, Merignac, France.

Vavilov, A. (1930) Wild progenitors of fruit trees in Turkistan and Caucasus. Int. Hort. Congress, London, UK, 271–286.

Voegt, E., and B. Goetz (1977) Weinbau. Ed. Ulmer, Stuttgart, Germany.

Waldin, M. (2004) Biodynamic wines. Beazley, London, UK.

Wilson, T.E. (1998) Terroir, geology/climate and culture of making French wines. Beazley, London, UK.

Winkler, A.J., J.A. Cook, W.M. Kliewer, and L. Lider (1974) General viticulture. Univ. of California Press, Berkeley, California.

APPENDIX 2

List of Tables, Figures and Illustrations

Tables

Number	title	page number
Table 1.1	Evolution of genus Vitis. (L.), after P. Galet.	2
Table 1.2	World distribution of viticulture (after Jackson & Schuster 2001).	5
Table 1.3	Selection of recommended varieties for different climatic zones (after Jackson & Shuster)	6
Table 2.1	Bud numbers retained under different pruning systems in high and low density plantings and canopy exposure to direct light under the different canopy systems (after Jackson).	23
Table 3.1	Stages of seasonal vine growth (after Baggiolini). (English translation from left to right)	46
Table 3.2	Example of foliar analyses report from Italy (after A. Paoletti).	57
Table 3.3	Bud fruitfulness along the cane (after Jackson & Schuster 2001).	63
Table 4.1	Berry growth and weight from fruit set to veraison (after Jackson & Schuster 2001).	65
Table 5.1	The varietal groups and multiplier used for each in estimating cluster weight at harvest.	84
Table 5.2	The long-term average periods of maturity in number of days from flowering to harvest for red grapes in Bordeaux and Burgundy (after Blouin & Guimberteaux, 2000).	85
Table 5.3	The maturation index of potential phenol content in grapes (after Glories)	86

List of Tables, Figures and Illustrations

Figures

Number	title	page number
Figure 2.1	Bordeaux cane pruned vine with or without spurs in winter.	26
Figure 2.2	Guyot pruned vine in winter.	27
Figure 2.3	Double Guyot, long cane with spurs and VSP canopy training in winter.	28
Figure 2.4	Bilateral cordon with 'Sylvoz', bent cane pruning and VSP canopy in winter.	29
Figure 2.5	Geneva double curtain vine in winter.	30
Figure 2.6	Scott Henry pruned vine in winter.	31
Figure 2.7	Quadrilateral, cordon-spur pruned vine in winter.	32
Figure 2.8	Mosel bent cane pruned vine in winter.	33
Figure 2.9	Bilateral Guyot cane with Lyre trellis, inclined canopy system in winter.	34
Figure 3.1	Primary bud, leaf, lateral and secondary buds in vines (after M. Fregoni 2001).	54

Photo illustrations

Number	title	page number
Cover Photo	Vineyard in the hills near Florence, Tuscany.	(front cover)
Illustration 1.1	Wild vine growing in a tree.	xvi
Illustration 1.2	Coast of ancient Thracia with Gobelet vines, Turkey.	3
Illustration 1.3 (Back cover)	Winter landscape in cool climate region.	7; (back cover)
Illustration 1.4	Summer landscape in hot climate region.	8
Illustration 1.5	Vine response to high soil salinity, Calabria.	9
Illustration 1.6	Terraced old Nerello vines on Etna, Sicily.	10
Illustration 1.7	Winter lime-sulphur spray application, Waiheke, NZ.	11
Illustration 1.8	Decomposing lava at 800m altitude. Etna, Sicily.	12
Illustration 1.9	Coastal clay loam soils with high level of salinity in Pakino region, Sicily.	13
Illustration 1.10	Alluvial brown loam soil over sandstone base rich in calcium carbonate, Avola region, Sicily.	14
Illustration 2.1	Root exploration of soil profile.	18
Illustration 2.2	Balanced VSP canopy.	21
Illustration 2.3	Gobelet vine (before pruning).	24
Illustration 2.4	Gobelet vine (after pruning).	24

Number	title	page number
Illustration 2.5	Gobelet vine in summer.	25
Illustration 2.6	Semi-dispersed canopy of Gobelet.	25
Illustration 2.7	Cane-pruned vine with VSP canopy in summer, Bordeaux.	26
Illustration 2.8	Guyot vine with VSP canopy in summer, New Zealand.	27
Illustration 2.9	Double Guyot, cane system with VSP canopy in summer, New Zealand.	28
Illustration 2.10	Bilateral cordon with 'Sylvoz' system after harvest, Marche.	29
Illustration 2.11	Geneva double curtain canopy system in summer, Friuli.	30
Illustration 2.12	Scott Henry trained canopy in summer, New Zealand.	31
Illustration 2.13	Quadrilateral cane with VSP canopy in summer, New Zealand.	33
Illustration 2.14	Mosel bent cane, stake trained vine canopy in summer, Germany.	34
Illustration 2.15	Bilateral Guyot cane with Lyre trellis and inclined canopy, France.	35
Illustration 2.16	Wild growing bush of American rootstock.	36
Illustrations 2.17 and 2.18	Large wounds after poor or re-constructive winter pruning.	41
Illustrations 2.19 and 2.20	Symptoms of eska decline, Tuscany (foliar and cane symptoms).	42
Illustration 2.21	Sanitation with latex paint containing fungicide, Napa Valley.	43
Illustration 2.22	Internal tissue damage due to severe pruning wounds to old wood.	43
Illustration 2.23	Winter pruning in Napa Valley, California.	44
Illustration 3.1	Spring bud burst, Tuscany.	45
Illustration 3.2	Basal bud push, due to retaining short spurs in winter, Montalcino.	47
Illustration 3.3	Poor growth due to excessive bud number left during winter pruning.	47
Illustration 3.4	Premature shoot thinning, Tuscany.	48
Illustration 3.5	Vines before well-timed shoot thinning, Tuscany.	48
Illustration 3.6	Vines after balanced shoot thinning, Tuscany.	49
Illustration 3.7	Spring frost damage.	50
Illustration 3.8	Well-supported spring VSP canopy, Tuscany.	50
Illustration 3.9	Spring canopy secured with wire clips, Tuscany.	51
Illustration 3.10	Foliar symptoms of 'spring fever' disorder, Sangiovese.	52
Illustration 3.11	Vines trimmed at flowering/early fruit set, Tuscany.	52
Illustration 3.12	Grape vine flower, Sangiovese.	53
Illustration 3.13	Combination of cluster/tendril indicating low bud fruitfulness.	56

List of Tables, Figures and Illustrations

Number	title	page number
Illustration 3.14	Poor set due to coloure disorder, Merlot.	58
Illustration 3.15	High-quality spring growth in Sangiovese.	59
Illustrations 3.16 and 3.17	Well supported VSP canopy of high quality canopy environment, Sangiovese.	60
Illustration 3.18	Overexposed grapes after excessive leaf removal in the fruit zone, Merlot.	61
Illustration 3.19	Symptoms of severe water stress, Merlot.	62
Illustration 4.1	Grape bunch collapse due to failure of the berry reflux mechanism.	68
Illustration 4.2	High altitude vines of Karazakis variety near ancient Troy, Turkey.	69
Illustration 4.3	Coastal vineyards of ancient Thracia, Asia Minor or modern Turkey.	69
Illustration 4.4	Poorly managed canopy in mid-summer, Montalcino.	71
Illustration 4.5	Green harvest in a Sangiovese vineyard.	72
Illustration 4.6	Dry farmed Gobelet vines in ancient Thracia, Turkey.	74
Illustrations 4.7 and 4.8	Mechanical leaf removal (before and after), Friuli.	75
Illustration 4.9	Vineyards on the coastal hills of ancient Lydia in modern Turkey.	77
Illustration 5.1	Evaluation of grapes for 'green harvest,' Cabernet sauvignon.	81
Illustration 5.2	Overcrowded grapes without 'green harvest,' Sangiovese.	82
Illustration 5.3	Overcrowded, leaf stripped grapes with botrytis, Merlot.	82
Illustration 5.4	Uniform cluster size and distribution after 'green harvest,' Syrah.	83
Illustration 5.5	Vines after mechanical harvest, Friuli.	88
Illustration 5.7	Mechanical pre-pruning of cordon-spur pruned vines, Tuscany.	90
Illustration 5.6	'Moser' adaptation of the high Sylvoz cordon with V-canopy in Napa Valley, California.	90
Illustration 5.8	Use of electric vineyard vehicles during hand harvest. Bolgheri, Tuscany.	90
Illustration 5.9	Mechanical harvest in Friuli.	91
Illustration 5.10	Transfer of grapes into gondola, Friuli.	91
Illustration 5.11	Guyot pruned Merlot vine, before pruning, Bolgheri.	92
Illustration 5.13	Well managed Sangiovese VSP canopy, Scansano in Maremma.	92
Illustration 5.12	Uniform bunch size and fruit ripeness of Sangiovese at harvest after earlier 'green harvest' (hard seed) Tuscany.	92
Illustration 5.15		94
Illustration 5.14	Cordon spur with elongation cane, Cabernet in Napa Valley.	94
Illustration 5.16	Scientific study of vine metabolic processes, Bolgheri-Italy.	95

Index

agrochemicals, 9, 10
420-A Millardet e de Grasset (*V. berlandieri* × *V. riparia*), 39
Ampelidaceae, 1
Anthracnose, 40
143-A or Aripa (*Aramon* × *V. riparia*), 39
ARG-1 or AXR, Ganzin No1 (*Aramon* × *V. rupestris*), 37
armillaria, 13
Aspergillus, 40
autumn
    mechanical or hand harvest, 87–88
    post-harvest vine management, 88–95
    yield estimates and grape harvest maturity indexes, 79–87

balanced shoot thinning, 49*f*
balanced vine, 19–20
balanced VSP canopy, 21*f*
Balkans, 9
Basal bud push, 47*f*
berry and seed growth, 65–66
berry color, 2
berry hydrology, 85
berry reflux mechanism, 68*f*
berry volume and weight, 73–74
bilateral, cane-pruned 'Scott Henry' system, 31–32
bilateral cane 'Lyre' trellis with inclined VSP canopy, 34–35
bilateral cordon with spurs, 29–30
bilateral Guyot cane, 34*f*
    with Lyre trellis, 34*f*
bio-diversity, 9
bio-dynamic grape growing, 14
bio-grow registrations, 12
black goo, 41, 42
41-B Millardet e de Grasset (Chasselas × Berlandieri), 40
Bolton, R., 85
Bordeaux bilateral cane system, 23–27
botryspheria, 13, 41
botrytis, 13, 40
brown, gray or black rots, 40
Brunello in Montalcino, 8
budburst and early shoot growth, 45–52
bud differentiation, 53
bud fruitfulness, flowering and fruit set, 43–58
bulk quality wines, 5
Busby, James, 4
bush pruning, 22

Cabernet, 2, 51
Cabernet sauvignon, 55
canopy construction, canopy quality and optimal size, 58–63
canopy/crop management, 8
carbohydrates, 70–73
    partitioning, 16
CCC (chlormequat), 55
Chardonnay, 11, 15, 54
Cladosporium, 40
climate zones, 4–7
    varieties, selection of recommended, 6*t*
coastal vineyards of ancient Thracia, 69*f*
coefficient of berry weight, 81
colonization, 4
Cortez, 4

Couderc 3309 (*V. riparia* × *V. rupestris*), 37
161-49 Couderc (*V. berlandieri* × *V. riparia*), 39
crop estimation, 79
crop levels, 70–73
Crown gall, 41
5C-Teleki (*V. berlandieri* × *V. riparia*), 39
cultivation of vines, 3

Diploidia, 40
dormancy, *see* winter
double guyot, long cane pruning, 28
drip system, 76
drought-resistant rootstocks, 19
dry farming, 74–77

early stem necrosis (ESN), 55
electric vineyard vehicles, 90*f*
enzymatic activity, 66
eska decline, 42*f*
eska disease, 42
Eutypa (*E. lata*), 13, 41

Fercal (*V. berlandieri* 333 E.M. × Colombard BC1), 37
fertilization, 10
foliar analyses, 57*t*
folletage, *see* water berry
forest plants, 1
fortified wines, 5
fruit maturity, 87
full-bodied red wines, 5

Geneva double curtain canopy, 29
Geneva double curtain vine, 30*f*
Georgia, 9
German 'Mosel' bent double cane, 'Gobelet' system, 33–34
Glories phenol maturity index, 86
Gobelet spur system, 22–23
Gobelet vine, 24*f*, 25*f*, 26*f*
grape growers, 3–4
grapegrowing and physiology of wine, 82*f*
grape maturation indexes, 84

grape varieties, 4
  climatic categories, 5
  commercial cultivation, 9
grape vine
  adaptations, 7–9
  domestication, 3
  flower, 53*f*
  vegetative means, 2
Gravesac (16-149 × Couderc 3309), 37
green harvest, 83, 83*f*
  grapes for, 81*f*
Grenache, 46
growth hormones in shoot tips, 16, 65
26-G (*Trollinger* × *V. riparia*), 39
Guyot cane system, 27–28, 27*f*, 28*f*

heat summation, 66
'hen and chicken' symptoms, 61
herbicides, 9
hybridization, 7

immunology, 9
industrial fertilizers, 9
inter-specific hybridization, 8
irrigation, 10, 67, 74–77
Italian and Spanish reds, 11
Italian 'eska,' 41

Kober 125 AA (*V. berlandieri* × *V. riparia*), 38
Kober 5BB (*V. berlandieri* × *V. riparia*), 37–38

labour intensive method, 80
late stem necrosis (LSN), 55
latitude and temperature adjusted index (LTI), 6
leaf growth, 46
leaf shape, 2
light bodied, aromatic whites wines, 5, 11
Lyre system, 34

Malbec, 49
malic acid, 66
Massilia (Marseilles), 4

maturation index of
    potential phenol, 86t
mechanical harvest, 91f
mechanical leaf removal, 75f
Merlot, 54
mildews, 41
101-14 Millardet (*V. riparia* × *V. rupestris*), 39
mosel bent cane, stake trained
    vine canopy, 34f
mosel bent cane pruned vine, 33f
Muscats, 11

Nero d'Avola, 46
Nielluccio in Corsica, 8
nitrate toxicity, 54
nitrogen-rich spring fertilization, 54

organic registrations, 12

pathogenic fungi, 13
  and bacteria, 40–44
1103-Paulsen (*V. berlandieri* × *V. rupestris*), 40
pesticides, 9
Petit verdot, 51
phenolic metabolism, 67
phenols, 70–73
phoenicians, 3
photosynthates, 55, 58
photosynthesis, 66, 70
phylloxera louse (*P. vastatrix*), 35
Pinot and Syrah in France, 8
Pinot gris, 11
Pinot noir, 11
Pinots, 15
Portalis or Riparia Gloire de
    Montpellier (*V. riparia*), 38
post-harvest irrigation, 89
premature shoot thinning, 48f
principal pruning and canopy training
    systems, 20–21
  bilateral, cane-pruned 'Scott Henry'
    system, 31–32
  bilateral cordon with spurs, 29–30
  Bordeaux bilateral cane
    system, 23–27

double guyot, long cane pruning,
    28
Gobelet spur system, 22–23
Guyot cane system, 27–28
quadrilateral cordon system, 30–31
progressive soil drying regime, 76
Prugnolo in Montelpuciano, 8
pruning, 20–22, 24f
  double guyot, long cane pruning, 2
    8
  quadrilateral cane or cordon-spur
    pruning system, 32
    bilateral cane 'Lyre' trellis with
      inclined VSP canopy, 34–35
    German 'Mosel' bent double cane,
      'Gobelet' system, 33–34
pruning systems, 8, 23t, *see also*
    principal pruning and canopy
    training systems

quadrilateral, cordon-spur
    pruned vine, 32f
quadrilateral cane or cordon-spur
    pruning system, 32
  bilateral cane 'Lyre' trellis with
    inclined VSP canopy, 34–35
  German 'Mosel' bent double cane,
    'Gobelet' system, 33–34
quadrilateral cane with
  VSP canopy, 33f
quadrilateral cordon system, 30–31

red wine consumption, 11
region-adapted forms, 8
region-specific development, 4
Rhône valley of France, 23
Richter 110-R (*V. berlandieri* × *V. rupestris*), 38
Riebeeck, Jan van, 4
Riesling, 2, 8, 11, 15, 46, 55
root
  division, 18
  exploration of soil profile, 18
  growth and metabolism, 17–18
  morphology and water/nutrient
    uptake, 16–20

rootstocks for soils and climates, 35–37
   420-A Millardet e de Grasset (*V. berlandieri* × *V. riparia*), 39
   143-A or Aripa (*Aramon* × *V. riparia*), 39
   ARG-1 or AXR, Ganzin No1 (*Aramon* × *V. rupestris*), 37
   41-B Millardet e de Grasset (*Chasselas* × *Berlandieri*), 40
   Couderc 3309 (*V. riparia* × *V. rupestris*), 37
   161-49 Couderc (*V. berlandieri* × *V. riparia*), 39
   5C-Teleki (*V. berlandieri* × *V. riparia*), 39
   Fercal (*V. berlandieri* 333 E.M. × Colombard BC1), 37
   Gravesac (16-149 × Couderc 3309), 37
   26-G (*Trollinger* × *V. riparia*), 39
   Kober 125 AA (*V. berlandieri* × *V. riparia*), 38
   Kober 5BB (*V. berlandieri* × *V. riparia*), 37–38
   101-14 Millardet (*V. riparia* × *V. rupestris*), 39
   1103-Paulsen (*V. berlandieri* × *V. rupestris*), 40
   Portalis or Riparia Gloire de Montpellier (*V. riparia*), 38
   Richter 110-R (*V. berlandieri* × *V. rupestris*), 38
   140-Ruggeri (*V. berlandieri* × *V. rupestris*), 40
   Rupestris St. George or Rupestris du Lot (*V. rupestris*), 38
   Schwarzmann (*V. riparia* × *V. rupestris*), 38
   SO-4 Oppenheim (*V. berlandieri* × *V. riparia*), 38
rudimentary cultivation, 2
140-Ruggeri (*V. berlandieri* × *V. rupestris*), 40
Rupestris St. George or Rupestris du Lot (*V. rupestris*), 38

Sangiovese, 46, 51
Sangiovese in Chianti, 8
Sangiovese in Italy, 8
sanitation with latex paint containing fungicide, 43*f*
Sauvignon blanc, 11
Schwarzmann (*V. riparia* × *V. rupestris*), 38
Scott Henry pruned vine, 31*f*
Scott Henry pruning system, 32
Scott Henry trained canopy, 31*f*
seedless hybrid varieties, 55
seedlings, selection, 2
self-fertile forms, 55
severe water stress, 62, 62*f*
shoots, morphology, 58
Smart, R., 32
soil
   density, 8
   fertility, 9, 75
   physical architecture, 17
   water conservation methods, 76
SO-4 Oppenheim (*V. berlandieri* × *V. riparia*), 38
spring
   budburst and early shoot growth, 45–52
   bud fruitfulness, flowering and fruit set, 43–58
   canopy construction, canopy quality and optimal size, 58–63
   fever, 51, 52*f*
   foliar fertilization, 51
   frost damage, 50*f*
stem necrosis, 55
sub-lateral shoot, 53
summer
   berry and seed growth, 65–66
   berry volume and weight, 73–74
   carbohydrates, 70–73
   crop levels, 70–73
   dry farming, 74–77
   irrigation, 74–77
   phenols, 70–73
   photosynthesis, 70
   respiration rates, water and nitrogen usage by vines, 66–69

vigor control, 74–77
Sylvoz bent canes, 29
Syrah, 49

temperate zone, 4
trade
   globalization of trade, 11
Trans-Caucasus, 9
Trichoderma species (*T. viridiae*), 13
Tuscany (foliar and cane
      symptoms), 42*f*

UC Davis, 5

varietal groups and multiplier, 84*t*
vertical positioned canopy (VSP), 59
vigneron, 11
vines
   adaptations, 9
   after mechanical harvest, 88*f*
   drought responses, 67
   growth, stages of, 46*f*
   primary bud, leaf, lateral and
      secondary buds, 54*f*
   respiration rates, water and nitrogen
      usage by, 66–69
   varieties, clones, 8
vine shoot elongation, 46
vines of Karazakis, 69*f*
vineyards, 3–4
   coastal hills of ancient Lydia, 77*f*
   fertilization of, 12
   industrial fertilizers in, 9
   soils, 17
viticulture
   bio-dynamic concept, 13
   grape vine adaptations, 7–9
   historical perspective, 1–7
   sustainable, 12
   systems, 10–14
   world distribution, 5*f*
Vitis. (L.), evolution, 2*f*
Vitis alba, 1
Vitis berlandieri, 7, 18, 19,
      35, 37–40, 67
Vitis siberica, 7
Vitis silvestris, 1

Vitis vinifera, 1, 4, 7, 8, 35, 36

water berry, 68
wild grape vine species, 1
wild wines (Vitis), 1
wine
   categories, 5
   consumption, 10
   industrial, 10
   quality, 3
wine, grapegrowing and
      physiology of, 82*f*
wine industry, 11
winter
   morphology and physiology of vine
      roots, 15–16
   pathogenic fungi and
      bacteria, 40–44
   principal pruning and canopy
      training systems, 20–21
      bilateral, cane-pruned 'scott henry'
         system, 31–32
      bilateral cordon with spurs, 29–30
      Bordeaux bilateral cane
         system, 23–27
      double guyot, long
         cane pruning, 28
      Gobelet spur system, 22–23
      Guyot cane system, 27–28
      quadrilateral cordon
         system, 30–31
      quadrilateral cane or cordon-spur
         pruning system, 32
      bilateral cane 'Lyre' trellis with
         inclined VSP canopy, 34–35
      German 'Mosel' bent double cane,
         'Gobelet' system, 33–34
   root morphology and water/nutrient
      uptake, 16–20
   rootstocks for soils and
      climates, 35–37
      420-A Millardet e de Grasset (*V.
         berlandieri* × *V. riparia*), 39
      143-A or Aripa (*Aramon* × *V.
         riparia*), 39
      ARG-1 or AXR, Ganzin No1
         (*Aramon* × *V. rupestris*), 37

winter *(Cont.)*
- 41-B Millardet e de Grasset (*Chasselas* × *Berlandieri*), 40
- Couderc 3309 (*V. riparia* × *V. rupestris*), 37
- 161-49 Couderc (*V. berlandieri* × *V. riparia*), 39
- 5C-Teleki (*V. berlandieri* × *V. riparia*), 39
- Fercal (*V. berlandieri* 333 E.M. × Colombard BC1), 37
- Gravesac (16-149 × Couderc 3309), 37
- 26-G (*Trollinger* × *V. riparia*), 39
- Kober 125 AA (*V. berlandieri* × *V. riparia*), 38
- Kober 5BB (*V. berlandieri* × *V. riparia*), 37–38
- 101-14 Millardet (*V. riparia* × *V. rupestris*), 39
- 1103-Paulsen (*V. berlandieri* × *V. rupestris*), 40
- Portalis or Riparia Gloire de Montpellier (*V. riparia*), 38
- Richter 110-R (*V. berlandieri* × *V. rupestris*), 38
- 140-Ruggeri (*V. berlandieri* × *V. rupestris*), 40
- Rupestris St. George or Rupestris du Lot (*V. rupestris*), 38
- Schwarzmann (*V. riparia* × *V. rupestris*), 38
- SO-4 Oppenheim (*V. berlandieri* × *V. riparia*), 38

vine dormancy, 15

Board and Bench Publishing

BOARDANDBENCH.COM

The Viticulture & Enology Library

Wine Faults
John Hudelson
$39.95 ISBN 978-1-934259-63-4

Acidity Management in Must & Wine
Volker Schneider
$45 978-1-935879-18-3

Practical Field Guide to Grape Growing & Vine Physiology
Schuster, Paoletti, Bernini
$45 ISBN 978-1-935879-31-2

View from the Vineyard
Clifford P. Ohmart
$34.95
ISBN 978-1935879909

Understanding Wine Technology
David Bird
$44.95
ISBN 978-1-934259-60-3

Concepts in Wine Technology
Yair Margalit
$45 ISBN 978-1-935879-80-0

Concepts in Wine Chemistry
Yair Margalit
$89.95 ISBN 978-1-935879-81-7

The Wine Business Library

The Business of Winemaking
Jeffrey L Lamy
$45 ISBN 978-1-935879-65-7

The Business of Sustainable Wine
Sandra Taylor
$45
ISBN 978-1-935879-30-5

Wine Marketing and Sales 3rd Ed.
Wagner, Olsen, Thach
$75 ISBN 978-1-935879-44-2

Artisan Public Relations
Paul Wagner
$29.95 ISBN 978-1-935879-29-9

How to Import Wine 2nd Ed.
Deborah M Gray
$29.95
ISBN 978-1-935879-40-4

Wine Business Case Studies
Pierre Mora, Editor
$35.00
ISBN 978-1-935879-71-8

Printed by Amazon Italia Logistica S.r.l.
Torrazza Piemonte (TO), Italy